THE HUMAN BODY

THE HUMAN BODY

Kenneth L. Jones
Louis W. Shainberg
Curtis O. Byer

MT. SAN ANTONIO COLLEGE

CANFIELD PRESS
SAN FRANCISCO

A Department of
Harper & Row, Publishers, Inc.

THE HUMAN BODY

Copyright © 1971 by Kenneth L. Jones, Louis W. Shainberg, and Curtis O. Byer

Printed in the United States of America. All rights reserved.

No part of this book may be used or reproduced in any manner whatsoever without written permission except in the case of brief quotations embodied in critical articles and reviews. For information address Harper & Row, Publishers, Inc., 49 East 33rd Street, New York, N.Y. 10016.

Standard Book Number 06-384364-1

LIBRARY OF CONGRESS CATALOG CARD NUMBER: 79-141238

CONTENTS

Preface   vii

1   THE SKELETAL SYSTEM   1

Parts — Formation and Growth of Bones — Movement of Bones — Summary — Questions

2   THE MUSCULAR SYSTEM   13

Structure of Skeletal Muscles — How Muscles Move the Bones — Kinds of Skeletal Muscles — Structure of Cardiac Muscle — Summary — Questions

3   THE NERVOUS SYSTEM   32

Structure of the Nervous System — Memory and Learning — Disorders of the Nervous System — Summary — Questions

4 THE EYES AND EARS   45

Structure of the Eye — Disorders of the Eye — Care of the Eyes — Structure of the Ear — Disorders of the Ear — Care of the Ears — Summary — Questions

5 THE SKIN AND HAIR   59

Structure of the Skin — Functions of the Skin — Hair — The Nails — Care of the Skin — Special Skin Problems — Care of the Hair — Summary — Questions

6 THE CIRCULATORY SYSTEM   74

The Heart — The General Circulatory Scheme — The Coronary Vessels — Physiology of the Heart — The Lymphatic System — Fetal Circulation — Blood — Blood Groups and Transfusions — Summary — Questions

7 THE RESPIRATORY SYSTEM   86

Respiration — Diffusion of Gases — Breathing — Capacity of the Lungs — Transport of Oxygen in Blood — Transport of Carbon Dioxide — Summary — Questions

8 THE DIGESTIVE SYSTEM   95

The Oral Cavity — Salivary Digestion — Gastric Digestion — Intestinal Digestion — Absorption — Summary — Questions

9 THE EXCRETORY SYSTEM   112

Materials to be Eliminated — The Urinary System — Formation of Urine — Urination — Summary — Questions

10 THE REPRODUCTIVE SYSTEMS   121

The Male Reproductive System — The Female Reproductive System — Menstrual Cycle — Pregnancy — Summary — Questions

11 ENDOCRINE GLANDS AND HORMONES   139

Glands and Hormones — The Endocrine Glands — Summary — Questions

Glossary   151

Bibliography   155

Index   157

PREFACE

Knowledge of the normal operation of the body is essential for each of us in order to achieve and maintain a maximum level of health. Much like a machine, the body consists of many special parts, each of which is formed and developed for a very specific task. The success of the operation of the whole machine depends upon the presence and efficient functioning of each essential part.

As exquisite as a modern automobile is, the human body is far more detailed. Its normal development and operation is so flawless that many people are unaware of its intricacy. It is possible for many to live normal-length lives, and a few to live exceptionally long lives, with little particular information about the body's operation. Although every man observes many points about his external structure, few people acquire detailed information on the internal aspects of their personal structure and function.

Such information is crucial. With knowledge of his body's systems, a person can avoid the half-truths of the quack. By understanding the body's structure and function, one can intelligently desist from conforming to the health practices of friends and independently follow those which are known to enhance a longer, healthier life. One acquires such freedom of action only when he knows the workings of his own body.

The Human Body is a systematic description, with illustrations, of each of the ten major systems, or parts, of the human body. The nervous system is subdivided in this book, with the eyes and ears treated separately in Chapter 4. The authors have carefully selected only those items of greatest importance and interest to the student. In the event that certain sections may appear abbreviated, it is because the authors have omitted material that they feel is too detailed for this particular book. Many excellent texts provide more complete information for the teacher and the inquisitive student. The authors have also omitted most of the effects of malformation and disease. Fuller treatment of the most important of the body systems can be found in other books in this series, the titles of which are listed on the back cover.

K.L.J.
L.W.S.
C.O.B.

THE
HUMAN
BODY

Chapter 1
THE SKELETAL SYSTEM

The skeletal system consists of all the bones and cartilage of the body, and the ligaments that hold joints together. The skeleton serves three major functions: protection, support, and, in association with the muscles, movement. It also serves as a reservoir for calcium and phosphorus, and as the location for the formation of red blood cells and some white blood cells.

PARTS

The skeleton consists of two subdivisions: the *axial skeleton*, the central axis of the body, and the *appendicular skeleton*, the bones of the shoulders, hips, arms, and legs (Figure 1.1).

The bones of the body may be classified in four general types in terms of their shapes: *long bones* (as in arms and legs), *short bones*

2 THE HUMAN BODY

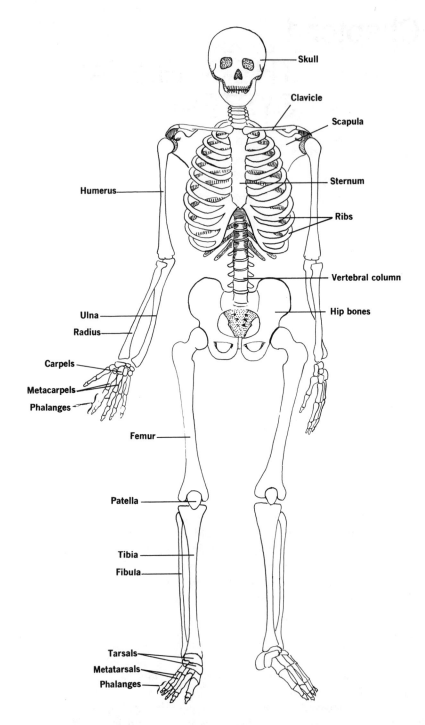

Figure 1.1 Human skeleton, front view.

(as in the wrists and ankles), *flat bones* (as in the skull and chest), and *irregular bones* (as in the vertebral column).

The adult skeleton consists of 206 bones. At birth, however, as many as 270 may be identified. The reduced number in the adult results from the union, or fusion, of a number of bones which are separate in the infant.

Axial Skeleton

The axial skeleton consists of the skull, vertebral column, ribs, and breastbone. In the adult this represents eighty bones.

SKULL

The skull (cranium and face) are formed by twenty-two bones. Many of these are flat bones and become inseparably joined along jagged edges or *sutures*. Several bones form the top of the skull: the *frontal bone*, which forms the forehead; the two *parietal bones*, which start at the top and form the two sides of the skull; the *occipital bone*, which forms the back of the skull; and the *temporal bones* found around and above each ear. The large hole in the back of the skull (the *foramen magnum*) is the opening through which the brain and spinal cord connect (Figure 1.2). During fetal development, the

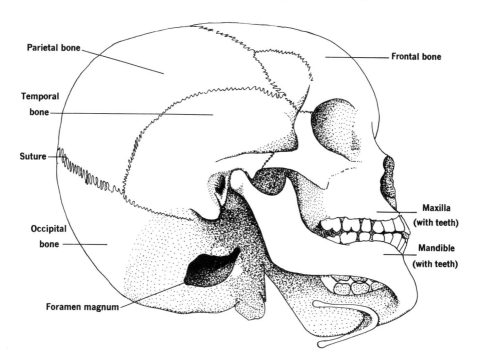

Figure 1.2 Skull, side view showing foramen magnum.

skull bones begin forming as widely separated plates. By the time of birth these plates have grown until they are touching each other in most places. The soft areas where they have not yet filled in are called *fontanels*. The fontanel, or soft spot, most commonly observed in the newborn is found near the top of the skull. Before long these fontanels disappear as the skull bones complete their development. Incomplete skull bone development at birth allows the infant's skull to be more pliable, and this permits an easier movement down the birth canal during delivery.

There are many bones making up the face—the eye sockets, the outside nose, and the jaws. These vary in shape and size depending on age, race, sex, and individual differences and will determine the contour of the face.

VERTEBRAL COLUMN

Also known as the backbone or spinal column, the vertebral column connects the shoulders to the hips, and supports the head. In the infant it consists of thirty-three irregular bones. In the adult several of these have fused, reducing the count to twenty-six. The *vertebrae* (plural; *vertebra*, singular) are constructed with a *body* to support the next vertebra, an *arch* to provide space for the spinal cord, and *processes* (spines) to give muscles something to attach to (Figure 1.3). Between each vertebra is a disk of cartilage (similar to the soft tissue in the end of the nose or the ear lobe) which helps to hold the vertebrae together and cushions them as a person moves.

The vertebral column is divided into five sections. The seven *cervical* vertebrae make up the neck, the twelve *thoracic* support the ribs, the five *lumbar* in the lower part of the back, the single *sacrum* connects to the pelvis, and the single *coccyx*, or tailbone.

CHEST CAGE

The chest cage consists of twelve pairs of ribs (you might try to count the twelve on each side of your chest). These are held together in front with a *sternum* (breastbone) which you can feel as you tap the middle of your chest.

Appendicular Skeleton

The appendicular skeleton consists of the two arms, the two legs, the shoulder bones (pectoral girdle), and the hip bones (pelvic girdle). It is made of 126 bones.

THE ARMS AND PECTORAL GIRDLE

As seen in Figure 1.1, there is a ring of bones making up the shoulders. The ring is known as the *pectoral girdle*, and it supports the arms. It consists of the two collar bones (*clavicles*) which

THE SKELETAL SYSTEM 5

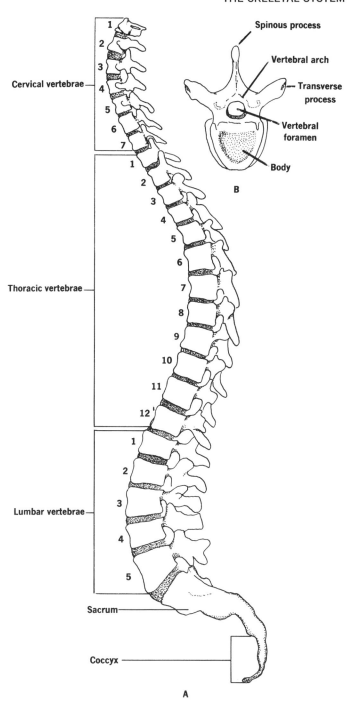

Figure 1.3 Vertebral column. (A) whole side view. (B) single vertebra, top view.

can be felt on the front of the shoulder, and the two shoulder blades (*scapulae*) which are deeply embedded in muscle on the back of the shoulders. The breastbone connects the two collar bones in the front.

Suspended from the ends of the shoulders are the arms. The upper arm (*humerus*) extends to the elbow. The lower arm consists of two bones. These are in a side by side position when the arm is held with the palm of the hand up. In this position the arm bone on the thumb side is the *radius;* the one on the little finger side is the *ulna*. The wrist is formed by eight small bones (*carpals*). The hand consists of the palm of the hand, the fingers, and the thumb. The five bones embedded in the muscle of the palm are the *metacarpals*. Each finger consists of three bones (*phalanges*), and the thumb consists of two. All of these bones of the arms and hands make possible a wide variety of delicate movements.

THE LEGS AND PELVIC GIRDLE

The *pelvic girdle* is made of two hip bones, one on each side, connected to the *sacrum* in the back and to each other in the front (front junction called *symphysis pubis*). Each hip bone is actually made up of three bones which have fused together. A man normally rests the belt of his trousers above the pelvic girdle and he rests his body on the lower part when sitting down.

The legs connect to the pelvic girdle. The upper leg (*femur*) extends to the knee. This is the longest and heaviest bone of the body. Protecting the knee is a knee cap (*patella*) which you can easily feel. The lower leg, or shank, consists of two bones. One is the shin bone (*tibia*) which can be easily felt on the front of the lower leg. The other (*fibula*) is smaller and deeply embedded in muscle. The ankle of the foot is made of seven *tarsals*. One of these forms the heel of the foot. The palm, or body of the foot, consists of five *metatarsals*. The big toe is constructed of two phalanges; toes two to five are each constructed of three phalanges. In order to support the weight of the body, the tarsals and metatarsals are formed into strong arches. Ligaments, tendons, and muscles help to maintain these arches.

FORMATION AND GROWTH OF BONES

The first skeleton formed in the developing embryo and fetus is made up of cartilage. Gradually the cartilage cells are replaced by bone cells. While this is going on, calcium is laid down around the cartilage and bone cells; a process called *calcification*. Then, the bone cells are organized in a definite fashion characteristic of bone tissue; a process called *ossification*. Ossification is far from complete in a child when born. In fact, it is believed to continue into a person's adult years. Although the adult skeleton is mostly bone, cartilage in certain parts of the body is never replaced by bone. Examples of this would be the flexible tip of the nose and the lobe of the ear.

The growth of bones can best be illustrated by the long bones of the body. Bones grow both in thickness and in length. The outside of these bones is covered by a tough membrane called *periosteum*. This membrane lays down bone cells and the diameter of the bone increases. Meanwhile, new bone cells are also laid down toward each end of the bone. First, new cartilage cells are being formed at the end of the bones, and then, these cells are replaced by bone cells. This occurs in an area of the long bones called the *epiphyseal plate* (Figure 1.4). There is always concern with a growing child when he breaks a bone near this plate. The fracture might do sufficient damage to the plate to slow down growth of that particular long bone. This could result in one arm or leg eventually being shorter than the other.

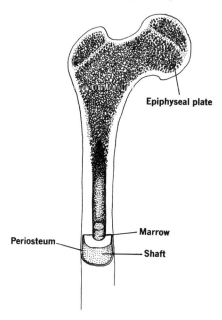

Figure 1.4 Bone, long section showing epiphyseal plate.

MOVEMENTS OF BONES

Bones are very much alive. The shape and composition of bones change with age and with the kinds of pressures they are subjected to. Bones tend to be elastic in young children, but as a person matures they become heavier and tougher. In old age bones become considerably lighter and more fragile. So, when a bone breaks, the speed of its healing will depend upon the type of fracture and the person's age.

Since the movement of bones is controlled by muscles, there must be a firm attachment of the muscle to the bone. Muscles attach indirectly to bones by *tendons* (Figure 1.5). The surfaces of bones

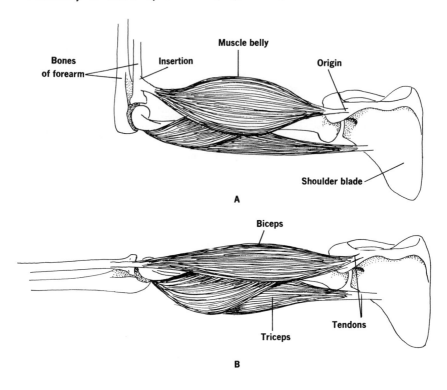

Figure 1.5 Muscles, insertion and origin on bones. Upper arm. (A) bended arm. (B) extended arm.

where tendons attach are usually roughened. This can be easily seen by looking at an arm or leg bone of a human skeleton. Muscles regulating the skeleton commonly attach to two different bones. Such muscles have an attachment end (*the origin*), and a movement end (the *insertion*). This might be made clearer by likening such a muscle to a boat cable used to pull water skiers. The end anchored to the boat would be the origin, and the action end the water skier holds onto would be the insertion.

Bones adapt their structure to the sort of stress they are placed under. Stress placed upon the skeletal muscles is transmitted to the bones. Strain on the leg bones applied over a long period of time causes them to become thicker. Consequently, bones of athletes are considerably thicker and heavier than those of nonathletes. When a particular part of a bone is subjected to continuing strain that part becomes thicker. The shape and thickness of bones reflect, then, the attachments of ligaments and muscle tendons, and the sorts of strains those muscles have exerted on the bones.

Joints

The region where one bone meets another is called a *joint*. Some joints are movable while others are not. Some kinds of joints in the body would be:

1. *No Movement.* Bone tissue forms between some joints, solidly fusing them together. The sacrum in the vertebral column was once five individual bones; the coccyx was once four. Some bones are tightly held together by fibrous tissue, such as the bones of the cranium (head) and face. Some of the cranial bones interlock with *sutures* (Figure 1.2).

2. *Slight Movement.* Some joints are held together by fibrous tissue in which there is a limited movement. Such fibers hold the vertebrae of the vertebral column together, but allow a limited movement in every direction (Figure 1.3).

3. *Free Movement.* Most of the joints of the body would fall into this category. The adjacent ends of the two bones are covered with cartilage. Tough connective tissues, or *ligaments*, hold one bone to the other. This joint is encased within a capsule which is filled with a fluid lubricant. This fluid helps to keep the ends of the bones slightly apart and cushions them. It also supplies nutrients to the tissues of the joints and removes wastes. There may also be other small fluid-filled sacs called *bursae*, where pressure is exerted over moving parts. When these sacs become inflamed, the condition is called *bursitis*. Freely moving joints may be further classified as: *gliding joints*, as in the wrist or ankle; *hinge joints*, as in the knee or elbow; and *ball and socket joints*, as in the hip and the shoulder.

Fractures

Fracture is a term applied to the breaking of a bone. A fracture may be either partial or complete. Bones of children have undergone less ossification and contain more organic material, and thus break less readily. With increasing age the bones consist of more inorganic material; they are more ossified and, therefore, are more brittle. They break easily and heal with more difficulty.

If bone is healthy, only unusual forces exerted in unusual directions can cause a fracture. Fractures of healthy bone due to accident are called *traumatic fractures*. A weakening of the bone from the effects of disease would be a *spontaneous fracture*. A fracture would be *incomplete* or *complete* depending upon whether the bone was partially or completely discontinuous. A fracture in which the bone protrudes through the outside tissue would be *compound*, or *open*. One not open to the outside tissue would be a *simple*, or *closed*, fracture.

Repair of Bone

There are three stages to the healing of fractures (Figure 1.6).

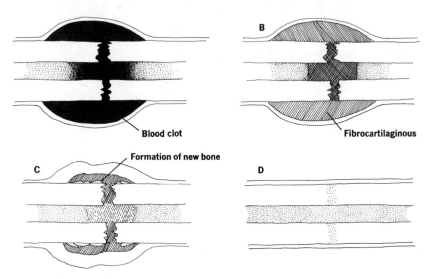

Figure 1.6 Healing of a bone fracture. (A) formation of procallus. (B) fibrocartilaginous callus. (C) bony callus. (D) completely healed fracture.

1. *Formation and organization of blood clot into a procallus* (Figure 1.6A). Accompanying a fracture would be the rupture of the blood vessels in the bone marrow and around the bone. Blood forms layers around the fracture and into the nearby tissues and clots from six to eight hours after the accident. Then, the blood clot is replaced by young connective tissue, a process occurring anywhere from one to four to seven weeks, depending on the size of the clot. This newly-forming connective tissue is called *procallus* (the first phase of new growth). The formation of the blood clot is essential to healing. A fracture will not heal if the clot is entirely removed. The procallus replaces dead tissues.

2. *Formation of fibrocartilaginous callus* (Figure 1.6B).

3. *Bony callus, or new bone, replaces the fibrocartilaginous material* (Figure 1.6C, D). The speed of healing of a bone is related to its location and the amount of *callus* produced by the body. For instance, fractures of the arms heal faster than those of the legs, and fractures in young people, who tend to produce more callus than older people, also heal faster. A *callus* is newly-formed material. Following

its beginning as procallus, the newly-formed material becomes *fibrocartilaginous* (something like the material in the end of the nose). Gradually, this is transformed into a bony callus, or growth. During callus formation, excess new bone is produced as a protective measure. After the fracture is again solidly reunited, the surplus bone around the place of the fracture will be gradually resorbed and the original shape of the bone will be reestablished.

SUMMARY
I. The Skeletal System
   A. Composed of all the bones of the body, cartilage, and ligaments that hold it together.
   B. Parts
      1. Axial skeleton—the central axis of the body.
      2. Appendicular skeleton—the bones of the shoulders, hips, and legs.
   C. Axial Skeleton
      1. Skull and face (formed by twenty-two bones).
      2. Vertebral column
         a. Also known as the backbone or spinal column.
         b. In an infant it consists of thirty-three bones. In the adult several of these bones fuse together reducing the number of bones to twenty-six.
      3. Chest cage consists of:
         a. Twelve pairs of ribs.
         b. Sternum (breastbone), which holds the ribs together in the front.
   D. Appendicular Skeleton
      1. The arms and the pectoral girdle—the ring of bones making up the shoulders.
      2. The legs and the pelvic girdle—made up of two hip bones connected to the sacrum in the back and to each other in the front.
II. Formation and Growth of Bones
   A. The skeleton is first formed in the developing embryo and fetus as cartilage which is gradually replaced by bone cells.
   B. Formation of bone tissue continues into the adult years.
III. Movement of Bones
   A. The movement of bones is controlled by muscles.
   B. Muscles regulating the movement of the skeleton commonly attach to two different bones.
IV. Joints—the region where one bone meets another.
   A. Some joints are movable while others are not.

B. The major kinds of joints are:
  1. No movement joints, where bone tissue has formed between joints fusing them together.
  2. Slight movement joints, where the joints are held together by fibrous tissue permitting only limited movement.
  3. Freely moving joints (most of the joints of the body). These include:
     a. Gliding joints of the wrist or ankle.
     b. Hinge joints of the knee or elbow.
     c. Ball and socket joints of the hip and shoulder.
V. Fractures
  A. Term applied to the breaking of a bone.
  B. If a bone is healthy, only unusual forces exerted in unusual directions can cause a fracture (traumatic fracture).
  C. Fracture due to a weakening of the bone from the effects of disease would be called spontaneous.
  D. Repair of bone
    1. There are three stages to the healing of fractures.
       a. The blood clot is formed and organized into a procallus.
       b. The procallus forms into a fibrocartilaginous callus.
       c. A bony callus, or new bone, replaces the fibrocartilaginous callus.
    2. The surplus bone around the place of the fracture will be gradually resorbed and the original shape of the bone will be reformed.

QUESTIONS FOR REVIEW
1. Explain the three major functions of the skeleton.
2. What kinds of tissue are part of the skeletal system? What are the functions of these tissues?
3. Differentiate between the axial and appendicular skeleton. Briefly explain the functions of these two subdivisions.
4. How do bones grow? How many years does it take for the skeleton to complete its growing processes?
5. Explain how muscles are attached to bone. Does the stress put on bones by muscles have anything to do with their growth?
6. What is a joint? Distinguish among the types of joints found in the human body.
7. Explain a fracture. Explain the differences between a compound and a simple fracture. In what ways may a bone be fractured?
8. Explain the process and stages in the healing of fractures.

Chapter 2
THE MUSCULAR SYSTEM

The ability to move the body is one of the most essential properties of human life. It is a keystone to the ability of a person to adapt to physical conditions around him. Movements also have esthetic values. There is fascination and mystique in the manner in which a person handles body movements.

All body movement depends on the activity of muscles. By definition, a *muscle* is an organ made up of bundles of contractile fibers by which movement is brought about.

There are three distinct kinds of muscle tissue found in the body: *skeletal*, which provides the force for the movement of bones; *smooth*, which is found in the walls of the digestive tract and blood vessels; and *cardiac*, which is found only in the heart. In this chapter, we will be concerned primarily with skeletal muscle.

STRUCTURE OF SKELETAL MUSCLES

It does not require technical training to tell that several sorts of tissue are present in skeletal muscle. Everyone has, on occasion, eaten beefsteak. It is easily seen that steak contains at least three sorts of tissue. As far as the animal is concerned, the fatty tissue is simply a reservoir for the storage of fat. The tough connective tissue, however, is all important. It surrounds the muscle fibers and holds them tightly together. It also joins with the tendon to connect the muscle securely to the bone so that movement can occur.

Each muscle consists of a main portion, called the *belly*, and two ends, which usually anchor to bones. As discussed in Chapter 1, one end is called the origin and the other the insertion.

All skeletal muscles consist of many muscle fibers. The larger the muscle the more fibers it contains. These fibers run parallel to the length, or long axis, of the muscle. When these fibers contract, the muscle contracts. Each muscle fiber is encased by connective tissue, something like a sausage inside its covering. Bundles of such fibers are, in turn, enclosed by connective tissue. Next, groups of muscle-fiber bundles are wrapped together by an outer connective tissue. Scattered through this connective tissue are nerves and blood vessels which supply the muscle fibers with nutrients and transmit messages as to when to contract. A cross section of a skeletal muscle showing the bundles of fibers is seen in Figure 2.1.

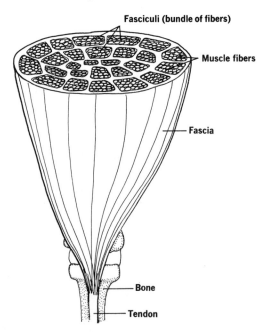

Figure 2.1 Section of a skeletal muscle showing muscle fibers.

Microscopic Structure

Each muscle fiber is a long cell containing many nuclei. Inside the cell is a semi-fluid called *sarcoplasm*. Within this sarcoplasm are found many long fibrils (little fibers) called *myofibrils*. These possess alternating light and dark bands around them. Around each muscle fiber is a membrane which is called the *sarcolemma*. The many nuclei inside of each muscle cell are usually located immediately under the sarcolemma. It is believed they are pushed there by the myofibrils when the fiber contracts. Although these muscle cells are microscopic in diameter (10 to 100 microns) they may be as long as the entire muscle (six to twelve inches long).

The light and dark shades of the bands which surround each myofibril are due to the presence of tiny filaments. When a skeletal muscle contracts these tiny filaments overlap more than usual. The overlap of filaments is greatest when the myofibril is in the contracted state and least in the relaxed states.

HOW MUSCLES MOVE THE BONES

Tone

Tone (also *tonus*) exists when a steady, partial contraction is maintained in a muscle. Muscle tone allows posture to be maintained for long periods of time with little or no fatigue. The absence of fatigue is brought about mainly by different groups of muscle fibers contracting in relays, giving each muscle group alternating periods of rest and activity. In man the greatest degree of tone is seen in the neck and back muscles. In an unconscious person the body collapses as if the muscles were completely relaxed. Even during sleep, tone exists, but it is at a minimum. In skeletal muscles tone provides firmness and creates a slight, steady pull on their bone attachments. Tone maintains a certain pressure on the abdominal wall, keeping the muscles of the abdomen firm. Tone is also exhibited in the walls of small arteries (maintaining blood pressure) and in the walls of the stomach and intestines (assuring the movement of materials through the digestive tract.)

Excitation (Stimulation)

A muscle may become active because the muscle fibers composing it become excited by a *stimulus*. If a cell, such as in a muscle fiber, is excited at any one place, the response will travel up or down the entire cell. Muscle fibers respond to a stimulus with a *contraction*. Muscles are normally stimulated by impulses carried to them by nerve cells (Figure 2.2), but they can also be stimulated artificially with electric current (as used in laboratories).

16 THE HUMAN BODY

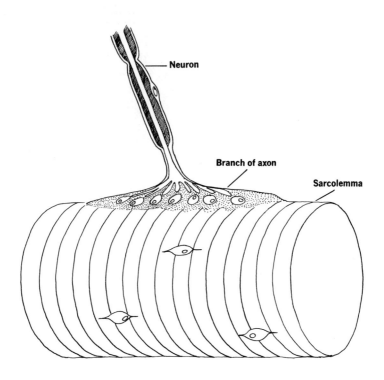

Figure 2.2 Attachment of nerve endings to a muscle cell. Nerve impulses travelling along nerve stimulate muscle cell to contract.

Muscle Contraction

Skeletal muscles contract quickly and relax promptly. When a stimulus arrives at the muscle, there are three steps to its reaction: a *latent period,* the time lag between the stimulus and the contraction; the *period of contraction,* the interval during which the muscle contracts and does work; and the *period of relaxation,* the time it takes for the muscle to return to the original length and relaxed state. How intense the contraction will be depends on several factors: (1) how strong the stimulus is, (2) how fast the stimulus is applied, (3) how long the stimulus lasts, (4) the weight of the load of work to be done, and (5) the temperature. There seems to be a middle point at which muscles do their best work. In man it is at normal body temperature (about 98.6° F.) under a moderate weight load, and under stimuli of moderate duration. Above or below this, muscles do not do their best work.

It has been found that any stimulus strong enough to cause a response in a muscle fiber will cause a *maximum* contraction, regardless of the strength of the stimulus. This is called the *all-or-none law*. Either a muscle cell contracts or it does not.

TWITCH

The contraction of a muscle in response to a single stimulus is called a *twitch*, and the muscle goes through the three reaction steps listed above. Such a single, isolated muscle twitch does not usually occur in a person's body under ordinary circumstances (perhaps with the exception of the eye-blinking movement).

Summation and Tetanus

When a muscle is stimulated to contract many times in succession up to a point, the contractions become more intense, creating a sort of staircase effect (Figure 2.3). If, while a muscle is at the peak of contraction, a second stimulus arrives, the maximum contraction will occur. This is called *summation of contractions*. If the successive stimuli arrive in such rapid succession that each occurs before the muscle can relax, the muscle will maintain a state of steady contraction, or *tetanus*. All voluntary acts of the body can be carried out because of tetanic contractions of the body muscles (such as holding the

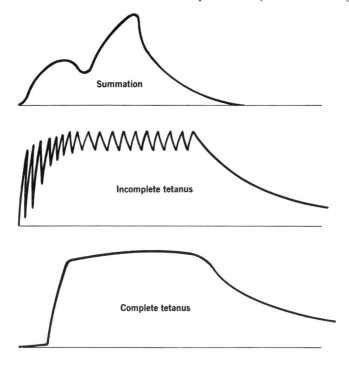

Figure 2.3 Effects of muscle contractions. Top, summation of contractions. Middle, incomplete tetanus. Bottom, complete tetanus. The higher the line, the stronger the muscle contraction.

body erect, or carrying a load). Even postural tonus is thought to be the result of a state of tetanus. This use of the word tetanus is not to be confused with the disease tetanus ("lockjaw"), in which there is a tetanus of the jaw and other muscles. With this condition it becomes impossible for the patient to relax these muscles at will.

Types of Contractions

When a weight is lifted, a muscle contracts and becomes shorter and thicker, but the tone remains the same. Since the tone of the cells is not altered, such contractions are called *isotonic*. If the muscle is made to contract against a weight it cannot lift, the tension in the muscle fibers increases, but the length of the muscle remains unchanged. Such contractions are called *isometric*. Most skeletal muscle contractions are isotonic.

Chemistry of Contraction

Energy is made available to the muscle cell in the form of *adenosine triphosphate* (ATP). This material is made through a process involving the breakdown of stored glycogen, which is a large carbohydrate molecule. Glycogen comes from the carbohydrates in our normal diet. The amount of ATP that can be produced from glycogen will depend on the amount of oxygen present in the body. Muscles operating with insufficient oxygen use up more glycogen and produce less ATP. When sufficient oxygen is present, the body can produce as much as ten to twelve times more energy from a given amount of glycogen than when the oxygen supply is insufficient.

With vigorous exercise, the body can use up more oxygen than is normally supplied to the cells. This creates an *oxygen debt*, which leads to the accumulation of certain wastes in the body, and causes fatigue. Then, the body demands rest.

HEAT FORMATION

Body heat is produced in several ways. Heat is liberated when a muscle cell becomes active, whether the muscle shortens or not. Heat is given off during the period of contraction, and during the period of relaxation.

Fatigue and Exercise

If a muscle is continuously stimulated, the muscle finally will refuse to respond. This is caused by an oxygen deficiency, and a build-up of carbon dioxide and other wastes. Another cause for fatigue is a deficient supply of nutrients to the muscle cell. When exercise is moderate, the body can eliminate wastes easily. But, if exercise is more intense, the body may need to rest to catch up on waste disposal and oxygen and nutrition replacement. Fatigue can also come from causes other than muscular. It can be associated with various emotional states, brought on through boredom or exhausting mental activity.

Exercise is beneficial to the body because it brings on a change in conditions for all cells throughout the body. Fresh blood is brought in and accumulated wastes are removed. Exercise increases the size, strength, and tone of muscle fibers. Massage may be used as a partial substitute for exercise. While exercise is desirable, using fatigued muscles may be injurious if the energy supply to the muscle cells becomes too low. Usually, however, the sensation of fatigue protects us from such extremes.

KINDS OF SKELETAL MUSCLES

Skeletal muscles constitute the "red flesh" of the body and account for 42 percent of the male's weight and 36 percent of the female's weight. There are over 400 skeletal muscles in the human body and although no attempt will be made to go over all of them, some of the more important ones will be mentioned.

Muscles of the Head and Neck

These muscles are concerned with mastication (chewing), facial expression, and movement of the head.

THE HEAD (Figure 2.4)

1. *Orbicularis oculi.* ' This is the circular muscle around each eyelid. One part closes the eyes gently as in dozing, the other closes them forcibly.

2. *Extrinsic ocular muscles.* There are six of these for each eye. They move the eyeballs.

3. *Orbicularis oris.* This muscle surrounds the mouth. When it contracts the lips are brought together; a more forceful contraction purses the lips, as when kissing or pouting.

4. *Masseter.* The masseter arises on the cheek bone and inserts into the lateral angle of the lower jaw. It raises the lower jaw, thus closing the mouth.

5. *Temporal (temporalis).* This muscle arises in a fan shape over the temporal bone and inserts on the lower jaw, and is used in chewing. It lies below the outermost muscles.

6. *Buccinator.* Found on the cheek, this muscle helps to keep food from escaping between the teeth while chewing. It becomes well developed in musicians who blow wind instruments.

7. *Tongue.* The highly specialized movements of the tongue are due to the action of four pairs of muscles within the tongue, the *intrinsic muscles,* and three pairs from below the tongue, the *extrinsic muscles.* The tongue is active in chewing, speaking, and swallowing.

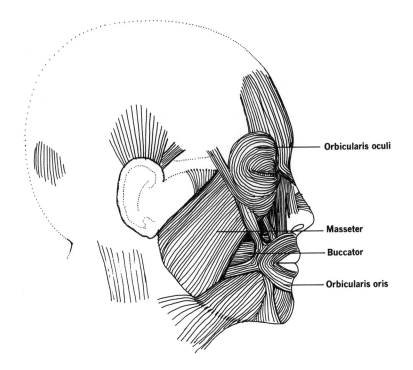

Figure 2.4 Removal of skin on face and head showing skeletal muscles.

THE NECK (Figures 2.5 and 2.6)

    1. *Sternocleidomastoid.* As the name may suggest, this muscle is attached to the sternum (breastbone), the clavicle (collar bone), and the mastoid process of the skull. It will pull the head toward the shoulder. The pair, one on either side of the head, help to keep the head erect.

    2. *Platysma.* Meaning "flat piece," this muscle arises from the skin of the chest, extends over the front of the neck and inserts on the lower edge of the lower jaw. It can pull the corners of the mouth and the lower lip downward, and, when contracted, can stretch the skin of the neck.

Muscles of the Back and Thorax

THE BACK (Figure 2.5 and 2.6)

    1. *Trapezius.* This is a large triangular muscle which arises on the back of the skull, fans down the back of the neck, and inserts into the shoulder bones. It can either raise or lower the shoulder or, if the shoulder is fixed, pull the head backward. It is important in maintaining posture.

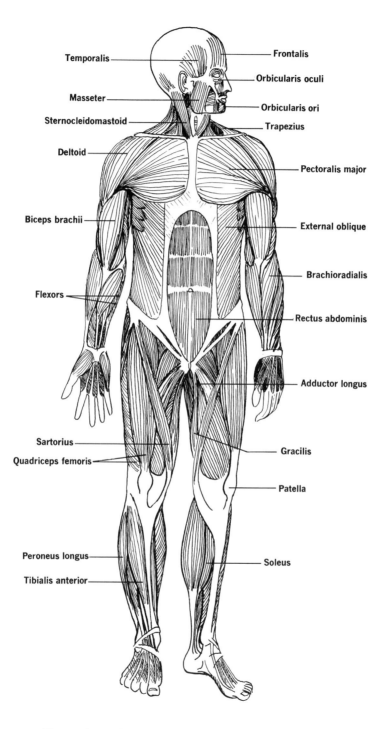

Figure 2.5 Muscles of the body, front view.

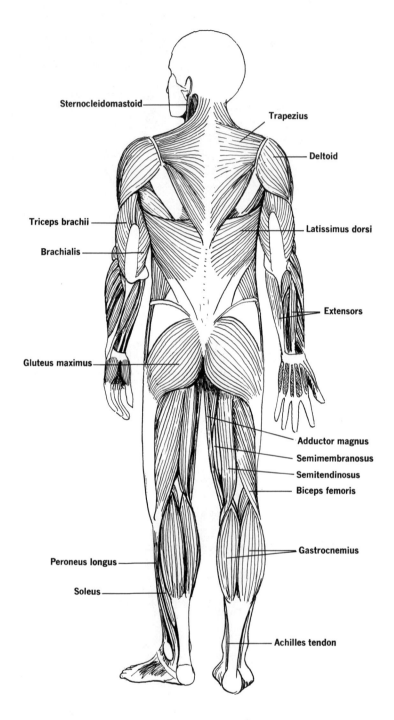

Figure 2.6 Muscles of the body, back view.

2. *Latissimus dorsi.* The latissimus dorsi is a very large, flat, triangular muscle, which covers the lower part of the back. It arises from the lower thoracic, the lumbar, vertebrae, and the sacrum. It sweeps upward and inserts on the humerus (upper arm bone). Upon contraction it will pull the arm and shoulder backward and rotate the arm inward. It may show very well on a muscular male.

3. *Rhomboids.* There is a pair of these muscles on each side of the body. They arise from the cervical and thoracic vertebrae and insert on the inner edge of the shoulder blade (scapula). Hidden under the trapezius; they move the scapula upward and inward.

4. *Deep back muscles.* There are several deep back muscles. The *sacrospinalis* is the most important. An elongated mass of muscle extending from the sacrum to the skull, it helps keep the back erect.

THE THORAX (Figure 2.5)

1. *Pectoralis major.* There are two pectoralis major muscles. Each is a large triangular muscle covering the upper front part of the chest. Each begins along the middle front of the chest, from along the entire clavicle and the sternum, and inserts onto the upper humerus. These important muscles assist in pulling the arms down toward the chest. They will also draw the arms across the chest while rotating them inward. These muscles are often well developed in swimmers.

2. *Pectoralis minor.* A small muscle entirely underneath the pectoralis major, this muscle arises on the upper margins of the third, fourth, and fifth ribs and inserts onto the outer tip of the scapula. It will draw the shoulder downward.

3. *Intercostals.* There are two of these muscles located between each pair of ribs. These muscles help to raise the chest during inhalation and depress it during forceful exhalation.

4. *Diaphragm.* This is a large dome-shaped muscle separating the thoracic and abdominal cavities, leaving three major openings between the cavities. The diaphragm is the principal muscle of inhalation. Upon contraction the dome pulls down and the space inside the chest is increased, allowing air from the outside to rush in. It also assists in expelling the fetus from the uterus, the feces from the rectum, the urine from the bladder, the contents from the stomach in vomiting.

Muscles of the Upper Extremities

THE SHOULDER (Figure 2.5 and 2.6)

Deltoid. The deltoid muscle forms the muscular cap of the shoulder. Arising on the scapula and clavicle, it moves down over the shoulder and inserts into the humerus. It is the chief muscle for moving the arm away from the body, and is also a common site for intramuscular injections.

THE ARM (Figures 2.5 and 2.6)

    1. *Biceps brachii.* More commonly referred to simply as the biceps, this is the large muscle on the front of the humerus which may bulge conspicuously in a well-developed person. (It is commonly displayed by small boys as proof of their strength.) The muscle arises at two different spots on the scapula (has two heads, or biceps), and inserts into the radius of the lower arm. This muscle pulls the arm forward and holds it close to the body. Thus it is important in lifting heavy objects.

    2. *Triceps brachii.* The triceps arises from three heads, two from the humerus and one from the scapula, and inserts on the ulna of the lower arm. It is sometimes referred to as the "boxer's muscle" since it is the major muscle for straightening the elbow. Thus it opposes the action of the biceps in that the biceps bends the elbow.

    3. *Brachialis.* Lying deep to the biceps, this muscle runs from the humerus to the ulna. It assists the biceps in flexing the elbow.

THE FOREARM (Figures 2.5 and 2.6)

    There are a number of small muscles in the forearm. These will be discussed in two groups: (1) those on the front of the arm (palm of the hand to front), and (2) those on the back of the arm (back of the hand to the front).

    1. *Front forearm muscles.* These muscles cause the wrist and fingers to pull up (or flex), and the hand to rotate so that the palm is turned down. These muscles include the *flexor carpi radialis, pronator teres,* and the *flexor carpi ulnaris.*

    2. *Back forearm muscles.* These muscles cause the straightening (extension) of the wrist and fingers and the rotation of the hand so that the palm is turned up. These muscles include the *extensor carpi radialis, extensor digitorium,* and *extensor carpi ulnaris.*

THE HAND

    Most of the muscles controlling the hand are lodged in the forearm and control the hand via long tendons (which can be felt on either side of the wrist). There are small muscles in the hand itself controlling movement of the fingers (separately or as a group) and the opposition of the thumb to the fingers.

Muscles of the Abdomen (Figure 2.5)

    The contraction of the abdominal muscles assists in such diverse things as: expiration of air (by forcing the diaphragm up), defecation, childbirth, urination, and maintaining the pressure in the abdomen. They can be conveniently grouped into three divisions.

1. *Lateral abdominal muscles.* The lateral abdominal wall is composed of three muscle layers. The fibers of each run in different directions, something on the order of crosslayers of plywood, thus providing the wall strength. The outermost is the *external oblique,* next deeper is the *internal oblique,* and the *transversus abdominis* is deepest.

2. *Anterior abdominal muscles.* The fibers of these muscles run vertically in a belt along the front center of the abdomen. The primary muscle here is the *rectus abdominis.*

3. *Posterior abdominal muscles.* There are several of these muscles beneath the latissimus dorsi and the sacrospinalis muscles. One of their functions is to help to keep a person from falling backward when standing.

Pelvic Floor

Two important muscles of the pelvic floor are the *external anal sphincter,* which controls the opening of the anus; and the *sphincter urethrae,* which controls the flow of urine through the urethra. Also stretching across the pelvic floor is the *levator ani,* which helps to retain the abdominal contents in position.

Muscles of the Lower Extremities

HIP AND BUTTOCK (Figure 2.6)

1. *Gluteus maximus.* This is the heaviest muscle of the body and forms most of the buttock. Arising from the lower vertebrae and pelvic girdle, it inserts on the femur. This muscle helps a person stand erect. Not only is it important in walking and running, but it is also a favorite site for intramuscular injections—particularly if the amount of medication is large.

2. *Gluteus medius and gluteus minimus.* Also making up the buttock are these two smaller muscles which help to rotate the femur inward. They lie beneath the gluteus maximus.

THIGH (Figures 2.5 and 2.6)

1. *Quadriceps femoris.* On the front of the thigh is a large muscle of four parts. This muscle arises on the femur and extends to the tibia of the lower leg. The patella (kneecap) is embedded in the tendon which connects this muscle to the lower leg. The quadriceps is the chief straightener (extensor) of the lower leg. This muscle extends the knee as in kicking a football. "Charley horse" is a term indicating a spasm, soreness, and stiffness in a muscle, and it is most commonly used in relation to this large thigh muscle. Occasionally injections are given here.

2. *Sartorius.* This is a long, narrow muscle that begins on the edge of the hip bone, winds downward and inward across the entire thigh, and ends up on the inner side of the lower leg. It is also called the "tailor's muscle" since it is chiefly used in crossing the legs. It once was the habit of tailors to sit in a cross-legged position.

3. *Biceps femoris.* This two-headed muscle arises on the femur and ilium and inserts on the lower leg. Along with the *semitendinosus* and the *semimembranosus,* the biceps femorus muscle makes up the hamstring muscles. These muscles serve chiefly to bend the knee.

4. *Adductor muscles.* The word *adduct* means to bring together. These muscles extend from the pubic bone (front of the body between the legs) to the femur. When one is lying on his back with his legs spread apart, these muscles are used to bring the legs together.

5. *Gracilis.* Arising from the pubic bone and inserting on the tibia, this muscle can be easily felt along the entire length of the inside of the thigh. This muscle becomes highly developed in those who ride horses.

THE LEG (Figures 2.5 and 2.6)

1. *Gastrocnemius.* This is the chief muscle of the calf of the leg and forms its curvature. It arises near the lower end of the femur and ends near the heel in a prominent cord on the back of the lower leg called the *Achilles tendon.* This largest tendon of the body attaches to the heel bone. The gastrocnemius (sometimes called the toe dancer's muscle) enables one to rise on the balls of his feet and toes, and helps bend the knee. The gastrocnemius of the frog is commonly used for laboratory experiments with muscle.

2. *Soleus.* Lying beneath the gastrocnemius, this muscle arises on the lower leg and attaches to the heel bone. It assists in extending the foot.

3. *Tibialis anterior.* This muscle arises on the upper part of the tibia and moves across the front of the leg to the inside of the ankle bones. It is used to turn the sole of the foot inward, and to raise the rest of the foot off the ground when one wants to walk on his heels.

4. *Peroneus longus and brevis.* These two peroneal muscles lie on the side of the leg. They turn the sole of the foot outward.

5. *Tibialis posterior.* This deep muscle on the back of the leg helps the support of the arch of the foot.

6. *Leg muscles that move toes.* Just as there were muscles in the arm controlling the fingers, so there are muscles in the lower leg, attached by long tendons over the top and along the bottom of the foot, to control the toes.

THE FOOT

Small muscles in the foot itself aid toe movement and help in the support of the arches. The toes and big toe do not possess the flexibility of the fingers and thumb.

STRUCTURE OF CARDIAC MUSCLE

Unlike skeletal muscle fibers which are distinct and separate, the fibers of cardiac muscle tissue fuse into one another. In skeletal tissue the nuclei are many and are confined to a single fiber; in cardiac tissue the nuclei do not seem to be distinctly related to a given fiber. Actually, the fibers of cardiac tissue do not appear to be organized in cells with distinct boundaries; they just seem to pass into one another.

Another distinction is in the way cardiac muscle responds to the "all-or-none" law. In skeletal muscle individual muscle cells follow the law, and the amount of contraction occurring in the muscle as a whole will depend upon the number of its muscle cells that contract. Cardiac muscle tissue also responds to the "all-or-none" law, but as an entire organ. All of the cells respond to the stimulus, not simply selected ones. Increasing the strength of the stimulus will cause no more intense a contraction, since all of the cells are already contracting to their capacity.

Impulses in the heart are not transmitted by nerves, as in skeletal muscle, but by specialized cardiac muscle cells.

Unlike skeletal muscle cells, cardiac cells do not regenerate. They do, however, have the ability to increase in size. When cardiac tissue is injured, as in a heart attack, scarring (forming of connective tissue) occurs by cell expansion. This is important to people with heart disease; but there are limits. If scarring occurs over a large area, the remaining functional part of the heart may be insufficient to pump enough blood to maintain normal activities, or even life.

SUMMARY

I. There are three distinct kinds of muscle tissue found in the body.
   A. Skeletal, which provides the force for the movement of bones.
   B. Smooth, found in the walls of the digestive tract and blood vessels.
   C. Cardiac, which is found only in the heart.
II. Structure of Skeletal Muscles.
   A. Each muscle consists of a main portion (the belly), and two ends, which usually anchor to bones.
   B. All skeletal muscles consist of many muscle fibers. The larger the muscle the more fibers it contains.

C. Microscopic Structure.
 1. Each muscle fiber is a long cell containing many nuclei.
 2. Inside the cell is found:
  a. a semi-fluid called sarcoplasm
  b. long fibrils called myofibrils
 3. Around each fiber (or cell) a membrane which is called the sarcolemma
III. How Muscles Move the Bones
 A. Tone (also tonus) exists when a steady, partial contraction is maintained in a muscle.
 B. Excitation (stimulation)
  1. Anything which brings about a muscle contraction is termed a stimulus.
  2. A muscle fiber can be excited anywhere along the cell.
 C. Muscle Contraction
  1. Skeletal muscles contract quickly and relax promptly.
  2. When a stimulus arrives at the muscle there will be three steps to a muscle contraction.
   a. A latent period, the time lag between the stimulus and the contraction.
   b. The period of contraction, the interval during which the muscle contracts and does work.
   c. The period of relaxation, the time it takes for the muscle to return to the original length and relaxed state.
  3. How intense the contraction will be depends upon several factors:
   a. how strong the stimulus is
   b. how fast the stimulus is applied
   c. how long the stimulus lasts
   d. the weight of the load of work to be done
   e. the temperature
 D. Twitch—contraction of a muscle in response to a single stimulus.
 E. Summation and Tetanus
  1. Contraction of muscles many times in succession, the contractions become more intense, creating a sort of staircase effect. If, while a muscle is at the peak of contraction, a second stimulus arrives, a maximum contraction will occur. This is termed summation of contractions.
  2. If successive stimuli arrive in such rapid succession that each occurs before the muscle can relax, the muscle will maintain a state of steady contraction, or tetanus.
 F. Types of Contractions
  1. Isotonic contractions occur when a muscle contracts and a weight is lifted.

THE MUSCULAR SYSTEM 29

      2. Isometric contractions occur when a muscle is made to contract against a weight it cannot lift or move.
  G. Chemistry of Contraction
      1. Energy is made available to the muscle cell in the form of adenosine triphosphate (ATP).
      2. More ATP is produced when large amounts of oxygen are present in the body.
      3. With vigorous exercise, the body can use up more oxygen than is supplied to the cells. This is called oxygen debt.
      4. Heat formation—heat is liberated when a muscle cell becomes active, whether the muscle shortens or not.
  H. Fatigue and Exercise
      1. Continuous contraction of a muscle will cause it to refuse to respond (or fatigue).
      2. During moderate exercise the body can eliminate the wastes that produce muscle fatigue.
      3. Fatigue can also be associated with various emotional states.
IV. Kinds of Skeletal Muscles—there are over 400 skeletal muscles in the human body.
  A. Muscles of the Head and Neck—mainly concerned with mastication (chewing), facial expression, and movement of the head.
  B. Muscles of the Back and Thorax
      1. The muscles of the back are important in:
         a. maintaining posture
         b. movement of the arms
         c. keeping the back erect
      2. The muscles of the thorax:
         a. pull the arms down and move them across the chest
         b. raise the chest during inhalation and depress it during forceful exhalation
  C. Muscles of the Upper Extremities
      1. Muscles of the shoulder form a muscular cap over the shoulder and move the arm away from the body.
      2. The upper arm muscles are important in lifting heavy objects.
      3. The forearm muscles:
         a. rotate the wrists and palms
         b. control the movement of the fingers
      4. The small hand muscles move the fingers (separately or as a group), and control the opposition of the thumb to the fingers.
  D. Muscles of the Abdomen
      1. Contraction of abdominal muscles assists in:
         a. expiration of air

b. defecation
c. childbirth
d. urination
e. maintaining the pressure in the abdomen
2. Posterior abdominal muscles help to keep a person from falling backward when standing.
E. Muscles of the Lower Extremities
1. Hip and buttock muscles help a person stand erect.
2. Thigh muscles extend the knee in kicking and walking, and help to spread the legs and bring them back together.
3. The leg muscles:
a. enable one to rise on the balls of his feet and toes
b. help bend the knee joint
c. extend the foot
d. lift the foot when one walks
e. help to support the arch of the foot
4. The muscles of the foot itself aid toe movement.
V. Structure of Cardiac Muscle
A. The structural differences between skeletal and cardiac muscles lie in the organization of the fibers and cells.
B. All cardiac muscle cells respond to a stimulus as a unit, causing the entire muscle mass to contract at once.
C. Nerve impulses in the heart are not transmitted by nerves, as in skeletal muscle, but by specialized cardiac muscle cells.
D. Cardiac muscle cells do not regenerate.

QUESTIONS FOR REVIEW

1. Make a diagram of a typical skeletal muscle and label its parts.
2. What are the distinguishing characteristics of the three types of muscle fibers in the human body?
3. Differentiate between tonus and tetanus.
4. What is the relationship between muscle tone and body posture?
5. Discuss the "all-or-none" law. Does the muscle as a whole operate according to this principle?
6. What controls the degree of contraction of a skeletal muscle?
7. Explain the steps involved in the contraction of a muscle. What takes place between the time a stimulus reaches a muscle until it returns to a relaxed state?
8. What is meant by the term "oxygen debt"? Explain the chemical mechanisms involved in muscle contraction and in muscle relaxation. What role do these processes play in causing fatigue?

9. Which muscles contract and which relax when you hold your right arm out at the side? When you throw a ball?
10. Which muscles are involved in the process of breathing?
11. What characterizes cardiac muscle? How does the "all-or-none" law function in cardiac muscle fibers?

Chapter 3
THE NERVOUS SYSTEM

The nervous system maintains at least some control over all other body systems and performs some rather unique functions of its own, such as sensory perception, thought, and memory. It is not surprising that the nervous system is the most complex and possibly least understood of all body systems. In this chapter, we will look at the structure and function of the nervous system.

STRUCTURE OF THE NERVOUS SYSTEM

The Neuron

The primary structural and functional unit of the entire nervous system is the nerve cell, or *neuron* (Figure 3.1). The number of neurons in the nervous system of a person has been estimated at over

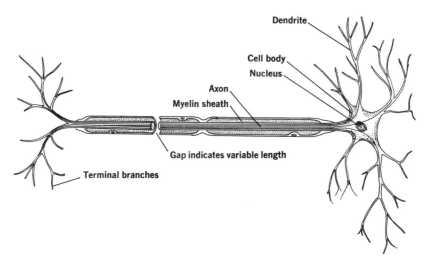

Figure 3.1 Neuron.

twelve billion. It is obvious that a neuron must be very small indeed. Every neuron has the same basic parts, although the relative sizes of these parts vary according to the specialized function of the neuron.

Each neuron has a swollen area called the *cell body*. Inside this cell body is the *nucleus* of the cell, which contains the genetic material of the cell and serves as the control center for the cell's activities. Compared with most other types of cells, a neuron has very little ability to repair damage. If the cell body is damaged, the entire nerve cell will die and never be replaced. Therefore, it is important to avoid any damage to the nervous system. Such damage may be permanent.

Extending out from the cell body are the impulse-carrying *nerve fibers*. Fibers carrying impulses toward the cell body are called *dendrites*. There may be one or several dendrites on each cell body. The fiber carrying impulses away from the cell body is the *axon*. There is only one axon per neuron.

Neurons are called *sensory neurons*, if they transmit impulses from sense organs to the brain or spinal cord; *motor neurons*, if they transmit impulses from the brain or spinal cord to an organ, such as a muscle or gland; or *connector neurons*, if they connect neurons to neurons.

A *nerve* is a bundle of nerve fibers. It may contain all sensory neurons, all motor neurons, or a mixture of the two. The connector neurons do not occur in nerves; they are found in the brain and spinal cord. The individual fibers in a nerve may be several feet long. For example, those fibers serving the fingers and toes extend all the way from the spinal cord.

TRANSMISSION OF NERVE IMPULSES

The transmission of an impulse by a nerve fiber is sometimes compared to the transmission of electricity along a wire. But actually, the nerve impulse involves electro-chemical changes in the neuron and travels much slower than the speed of electricity.

The junction between two nerve cells is called a *synapse* (Figure 3.2). In a synapse, the impulse reaching the end of the axon of one neuron stimulates an impulse to start down the dendrite of a second neuron. A nerve impulse can cross a synapse in one direction only, from axon to dendrite. Since nerve fibers can carry impulses in either direction, the synapse is important in eliminating wrong-way impulses.

Impulses are carried across a synapse by means of a chemical called *acetylcholine*. Acetylcholine is secreted by the end of the axon and causes an impulse to start down the dendrite of the second neuron. The acetylcholine is then rapidly destroyed by an enzyme

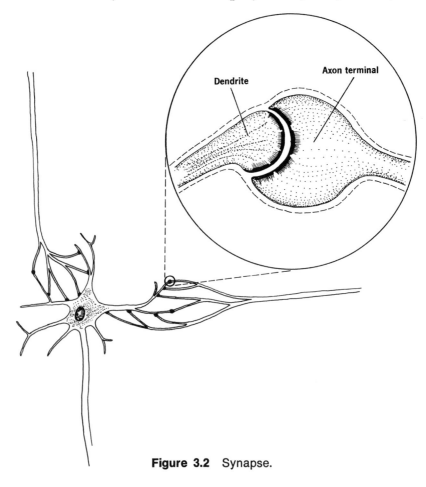

Figure 3.2 Synapse.

called *cholinesterase*. This destruction prevents the build up of acetylcholine in the synapse, which would result in a continuous stream of nerve impulses. The nerve gases of warfare, as well as certain common insecticides, act by inhibiting the enzyme cholinesterase. Acetylcholine can then build in the synapses, and the continuous nerve impulses that result can lead to death.

THE CENTRAL NERVOUS SYSTEM

The central nervous system consists of the *brain* and the *spinal cord*. The brain consists of three parts—the *cerebrum*, the *cerebellum*, and the *brain stem* (Figure 3.3).

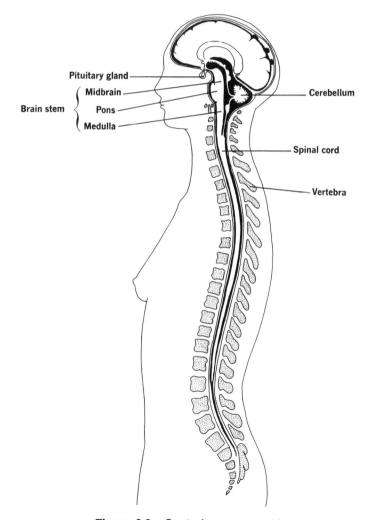

Figure 3.3 Central nervous system.

36 THE HUMAN BODY

The *cerebrum* is the largest and most highly developed part of the brain. It is the center of all conscious thought processes. It controls the voluntary muscles, interprets impulses from the sensory organs, and is the location for memory and learned behavior.

The cerebrum consists of two halves, the right and the left *cerebral hemispheres,* which are connected by *tracts* (bundles) of nerve fibers. Each hemisphere consists of four lobes — the *frontal, temporal, parietal,* and *occipital* (Figure 3.4). Specific areas of these lobes have been found to control specific body functions (Figure 3.5). In cases of brain damage, it is often possible to determine which part of the brain has been damaged by observing which functions are impaired.

The outer (surface) layer of the cerebral hemispheres, called the *cortex,* consists of *gray matter,* composed of the cell bodies of mil-

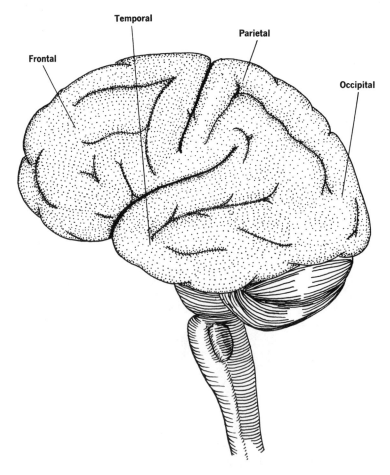

Figure 3.4 Lobes of the cerebrum.

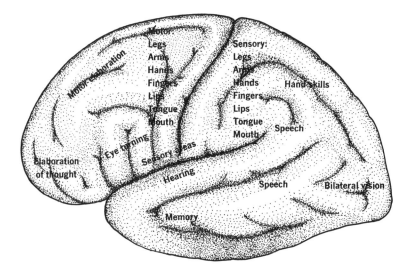

Figure 3.5 Cerebral function map.

lions of neurons. The inner portion of the cerebrum consists of *white matter*, composed of millions of neuron fibers, packed closely together.

The *cerebellum,* or "little brain," looks like a smaller version of the cerebrum. Like the cerebrum, it consists of two hemispheres and has a cortex layer of gray matter. The surfaces of both the cerebrum and the cerebellum are very wrinkled and folded into ridges. The purpose of this rough surface is to increase the surface area without increasing the total size of the brain. These ridges increase the amount of cortex and thereby of gray matter. The cerebellum is located at the rear base of the cerebrum. The function of the cerebellum is to coordinate muscle movements and to help maintain body balance and muscle tone.

The *brain stem* is a bundle of nerve tissues coming from the base of the cerebrum. It is directly continuous with the spinal cord at its lower end. The brain stem is the most primitive part of the brain. It is concerned with such basic animal functions as maintaining breathing, heartbeat, and digestive processes.

The *spinal cord* is a massive nerve column, continuous with the brain stem at its upper end, and extending downward through the *spinal canal* inside the vertebrae. It does not extend through all the vertebrae, but ends at the top of the second lumbar vertebra (just below the ribs).

The entire central nervous system is well protected. It is, first of all, enclosed in bone (the skull and vertebrae). And then, within this bone case, it is covered by three layers of membranes. These three membranes are called, collectively, the *meninges*. Between the second and third of these membranes, is a layer of fluid called the *cerebrospinal fluid*. This fluid layer varies in thickness, but is thick enough around the brain to act as a cushion or shock absorber, protecting the brain from minor jolts.

THE PERIPHERAL NERVOUS SYSTEM

The *peripheral nervous system* consists of the *nerves* connecting the brain and spinal cord to all the other parts of the body. There are twelve pairs of nerves arising from the brain, called *cranial nerves*. Most of these serve the organs of the head, such as the eyes and ears, although one pair (the vagus nerves) serves the internal organs of the chest and abdomen.

The nerves arising from the spinal cord are called *spinal nerves*. There are thirty-one pairs of these, one nerve arising from each side of each vertebra. Each spinal nerve emerges from the vertebra as a *dorsal root*, containing sensory neurons, and a *ventral root* containing motor neurons (Figure 3.6). These two roots join together to form the spinal nerve. The spinal nerves serve all the parts of the body below the head.

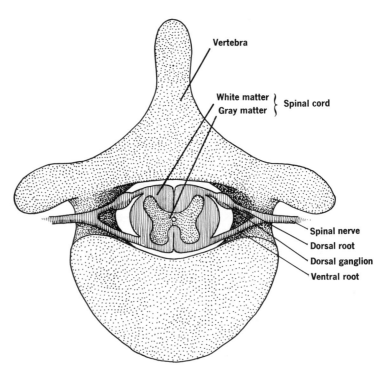

Figure 3.6 Spinal nerve emerging from spinal cord.

Reflexes

A *reflex* is an automatic response to a certain stimulus. It is carried out very rapidly, without any conscious action, and the same stimulus will always produce the same reflex action. A well-known reflex action is the knee jerk produced by tapping a tendon just below the kneecap. The neuron pathway producing this reflex is shown in Figure 3.7. There are hundreds of other human reflexes. For example, the contraction of the pupil of the eye in response to bright light is a reflex action, as is the jerking back of a hand which touches a hot surface. The hand is withdrawn even before the pain is consciously felt. The reflex action is carried by neurons within the spinal cord, while an impulse must reach the brain in order for the pain to be felt.

The Autonomic Nervous System

The *autonomic nervous system* regulates the functions of the various internal organs of the body, such as breathing, digestion,

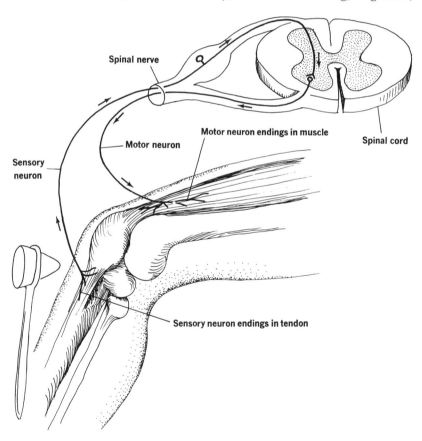

Figure 3.7 Reflex arc.

heartbeat, blood pressure, and glandular secretion. All action of the autonomic nervous system is subconscious and involuntary. The control of the heartbeat, for example, requires no conscious thought; nor is any conscious control usually possible.

The autonomic nervous system is not actually separate; it is a part of the peripheral nervous system. The autonomic nerves come from the brain and spinal cord.

There are two divisions of the autonomic nervous system, the *sympathetic* and the *parasympathetic*. These two divisions generally act in direct contrast to each other. The sympathetic division acts to prepare the body for emergency action. It is active during periods of fear, anger, or stress. The effects of the parasympathetic division are generally to restore and conserve energy. Some of the effects of the two autonomic divisions are compared in Table 3.1.

Table 3.1 The Autonomic Nervous System

PART OF BODY	EFFECT OF SYMPATHETIC DIVISION	EFFECT OF PARASYMPATHETIC DIVISION
Heart	Stimulates beat	Inhibits beat
Blood vessels	Dilates vessels in muscles, contracts others (raises blood pressure, forces more blood to muscles)	Dilates vessels in digestive tract, contracts others
Digestive process	Inhibits	Stimulates
Pupil of eye	Dilates	Constricts
Bronchi of lungs	Dilates	Constricts
Sweat glands	Stimulates	No action
Adrenal gland	Stimulates	No action

Although valuable during short periods of emergency, as when a person is under prolonged emotional stress, the effects of the sympathetic division can be harmful and lead to many *psychosomatic disorders*. These are physical conditions which are caused or aggravated by emotional stress. Some of the more common psychosomatic disorders include indigestion, digestive ulcers, asthma, high blood pressure, skin rashes, headaches, and menstrual problems. It should be stressed that each of these disorders can also have causes entirely independent of the emotions. But each one *may* be psychosomatic.

As an example of how psychosomatic disorders may be caused, let us consider one of the most common of these—digestive (peptic) ulcers. Under stimulation of the sympathetic division of the autonomic nervous system (in times of stress), the production of most of the digestive juices is stopped. This allows the body to put its full capacity toward immediate emergency action such as fighting or running from danger. But the production of stomach acid continues after the production of the juice which is supposed to neutralize this acid has been stopped. Thus, the stomach acid accumulates until it becomes strong enough to eat a hole in the wall of the stomach or small intestine. That hole is an ulcer. A person with a very resistant digestive tract might never develop ulcers, regardless of great emotional stress. Another person with a very sensitive digestive tract, might develop an ulcer even though he was subject to very little stress.

MEMORY AND LEARNING

The nature of memory and learning has been one of the most elusive aspects of human physiology. Only in recent years has some understanding of the process of information storage and retrieval been attained.

At least two types of memory are believed to exist. One is temporary and depends on the prolonged excitation of the involved neurons. Another is long-lasting and apparently depends on physical and/or chemical changes in the synapses between neurons. We know that prolonged memory cannot depend on continued activity of the nervous system. The brain can be totally inactivated, by such methods as cooling, anesthesia, or oxygen shortage, so that all function ceases; yet stored memories are retained when normal functioning is restored.

Several theories have been proposed to explain long-term memory. One is that the number of synaptic connections is increased in the learning process. Another holds that a chemical, ribonucleic acid (RNA), active in protein synthesis, is accumulated in the storage of information.

A prolonged memory is not established in the central nervous system in the first few minutes after a sensory stimulus. Rather, an hour or more is required to "fix" the memory. If many strong stimuli are received one after another, it usually will not be possible to remember any one of them very well. Each successive stimulus apparently disrupts the short-term memory circuit of the previous stimulus before its permanent memory can be established.

DISORDERS OF THE NERVOUS SYSTEM

Most of the disorders of the nervous system seem to fit into one of two basic categories, though the placement of many disorders is still uncertain. These two categories are:

1. *Emotional disorders.* These are *functional* disorders in which there is impairment in one's ability to adapt smoothly and effectively to stresses. There is no apparent organic damage to the nervous system. In some cases, chemical differences have been shown in the brains of persons suffering from emotional disorders. But, it has not been clearly established whether these differences are the cause of the problem or the result. If it can be shown that a chemical change is the *cause* of a problem, then that problem would fit into the second category.

2. *Neurological disorders.* These are *organic* disorders in which there is definite damage to the nervous system or abnormality in its chemical makeup. Any behavorial abnormality, or other sign or symptom, is the result of this basic disorder of the nervous system.

The emotional disorders are the concern of psychologists and psychiatrists. They are commonly treated with *psychotherapy* or mood altering drugs. The neurological disorders are the concern of neurologists, who must use many approaches to their treatment, often including surgery and/or drugs.

SUMMARY

I. Structure of the Nervous System
   A. The neuron (nerve cell) is the basic unit of the nervous system.
   B. Each neuron has several parts.
      1. Cell body—contains the nucleus.
      2. Dendrites—fibers carrying impulses toward the cell body.
      3. Axon—a single fiber carrying impulses away from the cell body.
   C. Neurons are classified as:
      1. Sensory—if they transmit impulses from sense organs to the brain or spinal cord.
      2. Motor—if they transmit impulses from the brain or spinal cord to an organ, such as a muscle or gland.
      3. Connector—if they connect neurons to neurons.
II. Transmission of Nerve Impulses
   A. A nerve impulse is a wave of electro-chemical changes traveling along a neuron.
   B. The junction between two nerve cells is called a synapse.
   C. Impulses are transmitted across a synapse by means of a chemical called acetylcholine.

III. The Central Nervous System
   A. Consists of the brain and spinal cord.
   B. The brain consists of three parts.
      1. The cerebrum
         a. Largest and most highly developed part.
         b. Center for all conscious thought processes.
         c. Consists of two halves, the right and left cerebral hemispheres.
         d. Each hemisphere consists of four lobes—frontal, temporal, parietal, and occipital.
         e. Outer layer, called the cortex, is gray matter, composed of cell bodies.
         f. Inner portion is white matter, composed of neuron fibers.
      2. Cerebellum
         a. Looks like a smaller version of the cerebrum.
         b. Coordinates muscle movements, maintains body balance and muscle tone.
      3. Brain stem
         a. Is directly continuous with the spinal cord.
         b. Most primitive part of brain.
         c. Governs breathing, heartbeat, and digestive processes.
   C. Spinal cord is a nerve column extending downward through the vertebrae.
   D. Entire central nervous system is well protected.
      1. Enclosed in bone (skull and vertebrae).
      2. Covered by three layers of membranes collectively called the meninges.
      3. Circulating between the second and third layers of the meninges is a layer of fluid called the cerebrospinal fluid.
IV. The Peripheral Nervous System
   A. Consists of the nerves connecting the brain and spinal cord to the rest of the body.
      1. Nerves arising from the brain are called cranial nerves.
      2. Nerves arising from the spinal cord are spinal nerves.
   B. Reflexes—are automatic, unlearned responses to certain stimuli.
   C. The Autonomic Nervous System
      1. Regulates the functioning of all internal organs.
      2. Consists of two divisions.
         a. Sympathetic—prepares the body for emergency actions.
         b. Parasympathetic—acts to restore and conserve energy.
V. Memory and Learning
   A. Two types of memory are believed to exist.
      1. One is temporary and depends upon the prolonged excitation of the involved neurons.

2. Another is long-lasting and apparently depends upon physical and/or chemical changes in the brain.
    B. Learning (or permanent memory) is still subject to a number of theories.
        1. One is that the number of synaptic connections is increased in the learning process.
        2. Another holds that a chemical is accumulated in the storage of information.
VI. Disorders of the Nervous System—fit into one or the other of two basic categories.
    A. Emotional disorders
        1. Impairment in one's ability to adapt smoothly and effectively to stresses.
        2. There is no apparent organic damage to the nervous system.
    B. Neurological disorders are organic disorders in which there is definite damage to the nervous system.

QUESTIONS FOR REVIEW

1. What is the basic structural unit of the nervous system? What are its key parts?
2. Distinguish among sensory neurons, motor neurons, and connector neurons.
3. What is a neuron? What is the difference between a neuron and a nerve?
4. What is the function of acetylcholine? Is this material used throughout the nervous system?
5. Name and briefly describe the parts of the central nervous system. What are the basic functions of each of these parts?
6. What is a reflex? What control do you have over the reflexes of your body?
7. What is the function of the autonomic nervous system? Name the two divisions of the autonomic nervous system. What are their functions?
8. What are psychosomatic disorders? What part does the nervous system play in psychosomatic disorders? What are some of the more common psychosomatic disorders?
9. Many theories have been proposed to explain memory. What are two of the more common theories of memory formation?
10. Explain the differences between emotional disorders and neurological disorders. What specialists are commonly called upon to treat these disorders?

Chapter 4
THE EYES AND EARS

The eyes and ears are man's primary sources of information and contact with reality. The loss of either one of these senses seriously handicaps a person. Much blindness and deafness can be prevented if something is known about the eyes and ears and their care.

STRUCTURE OF THE EYE

The eye (Figure 4.1) converts the energy contained in light waves into nerve impulses, which are sent to the brain. In order to see sharply, the light rays must be focused (bent) so that the rays coming from a single point on the object in view fall on a single point on the light-sensitive part of the eye. If the rays from a single point are scattered over the light-sensitive surface, blurry vision will result.

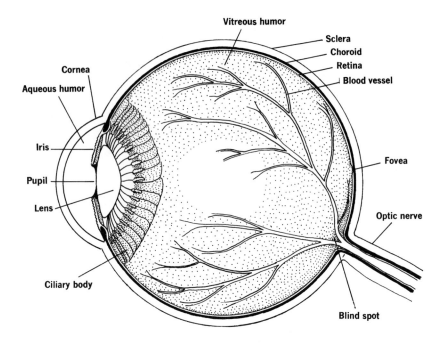

Figure 4.1 Cross section of the eye.

The focusing of light rays is carried out by two of the eye's structures. The first of these is the *cornea*, the clear "window" on the front of the eye through which light enters. The curved surface of the cornea bends the light rays, giving a "rough" focus. The focus is then refined by the *crystalline lens*, usually just called the *lens*. The lens should be perfectly clear and somewhat flexible. It is the lens that enables the eye to *accommodate* (to adjust for) light coming in from different distances. When a distant object is viewed, the lens assumes a flattened shape; for closer objects, it becomes more oval. The shape of the lens is controlled by the *ciliary body* which encircles the lens and is connected to it by the *suspensory ligament*.

The space between the cornea and the lens is filled with a watery fluid called the *aqueous humor*. The larger chamber behind the lens is filled with the jelly-like *vitreous humor*.

Just in front of the lens is the pigmented *iris*. This is the part of the eye that is brown, blue, hazel, and so forth. The opening in the center of the iris is the *pupil*. The size of the pupil is adjusted by reflex action by muscles in the iris, according to the light intensity.

The wall of the eyeball consists of three distinct layers of tissue. The outermost layer, called the *sclera*, is thick and rigid, giving the eyeball its shape. The sclera is seen from the front as the white of the eye. The cornea is a continuation of the sclera.

The middle layer is the *choroid,* a thin, dark-brown membrane lining the inner surface of the sclera. The choroid possesses many blood vessels which supply the innermost layer of the eye, the retina. The iris and ciliary body are continuations of the choroid.

The innermost layer of the eyeball is the *retina,* the light-sensitive part of the eye. It is a thin gray layer consisting of *millions* of specialized neurons. These light-receptor neurons are of two types, the *rods* and the *cones* (Figure 4.2). The rods are sensitive only to black, white, and shades of gray. The cones are sensitive to colors. The rods can function in much dimmer light than the cones, so twilight vision is in black and white, with the colors absent. The nerve fibers from the retina unite to form the *optic nerve* which carries the impulses to the brain. Since there are no rods or cones at the point where the retina joins the optic nerve, there is a true *blind spot* there. Just above the blind spot, in the exact center of the retina, is a tiny pit called the

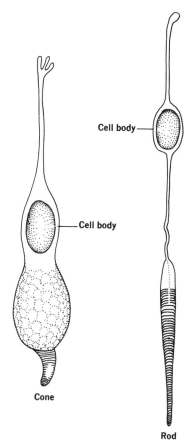

Figure 4.2 Greatly enlarged cone and rod from the retina of the eye. The retina contains millions of these modified neurons.

48 THE HUMAN BODY

fovea. The fovea consists entirely of cones; it contains no rods. This is the area of sharpest vision (acuity), and in reading or any intense vision, this is the most active part of the retina.

Near the fovea, the retina contains both rods and cones. As the distance from the fovea increases, the rods gradually become more abundant than the cones. The edges of the retina contain only rods, no cones. Any object seen out of the extreme "corner" of the eye, will fall on the edge of the retina and be seen in black and white only.

Lining the eyelids and covering the front of the sclera is a delicate membrane called the *conjunctiva*. This membrane forms a blind pocket around the eye which prevents foreign objects from getting around behind the eyeball. The conjunctiva is kept moist by the constant flow of tears from the *tear glands*. Tear fluid lubricates the eyelids, washes away foreign particles, and inhibits the growth of many bacteria. Excess tear fluid drains through small ducts into the nasal cavity.

DISORDERS OF THE EYE

Since anything that reduces the effectiveness of the eyes can be a serious handicap in one's everyday life, it is important that any eye defect receive prompt professional treatment.

Focusing Defects

If the light rays reflected from a given point do not all fall onto the retina at the same point, blurry vision will result. This problem can come about in several ways.

1. *Myopia (nearsightedness)*. In myopia (Figure 4.3) the eye is unable to clearly focus on distant objects. The light rays from distant objects come to their focal point somewhere in front of the retina, then cross, producing a blurry image when they reach the retina. This often hereditary condition is corrected by wearing lenses which throw the point of focus back onto the retina.

2. *Hyperopia (farsightedness)*. Hyperopia is the opposite of myopia. Distance vision may be satisfactory, but light rays from close objects are focused at some point behind the retina, again producing a blurred image. Sometimes, the muscles of the ciliary body may be able to compensate by thickening the lens, but this extra muscle effort may lead to painful eyestrain after long sessions of close work or reading. Like myopia, hyperopia is easily corrected with lenses (such as those found in reading glasses).

3. *Astigmatism*. Astigmatism is a mixed vision, in which the light rays do not all come to focus on the same plane. Some may focus in front of the retina, while others are focused at a point behind the retina. As a result, vision is blurred or distorted to some degree,

THE EYES AND EARS 49

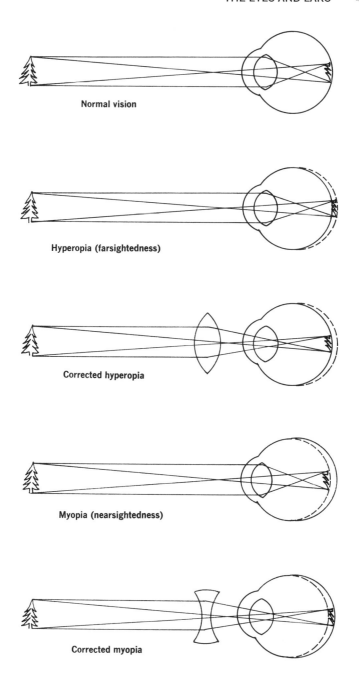

Figure 4.3 Normal and defective focusing. In the normal eye, light is focused exactly on the retina. In the farsighted (hyperopic) eye, light focuses behind the retina until corrected by a convex glass lens. In the nearsighted (myopic) eye, light focuses in front of the retina until corrected by a concave lens.

and the eyes are strained from trying to compensate for this. Astigmatism is due to an irregularity in the curvature of either the cornea or the lens. It is corrected by specially ground lenses.

4. *Presbyopia.* This is a type of farsightedness which often develops with increasing age. The lens loses its elasticity and becomes fixed in the position for distance vision. Close objects are blurred. Presbyopia is easily corrected with prescribed lenses.

Muscle Imbalance

Each eyeball is turned in its socket by six muscles. Normally the turning of the eyes is coordinated so that they both look at the same place. But sometimes the muscles of the two eyes are unequal in length or strength. As a result, one eye may be turned in, out, up, or down. This condition is called *strabismus.* The brain often learns to ignore the image coming from one eye, so the person really just sees with one eye at a time. Strabismus can often be corrected by an ophthalmologist (physician specializing in the correction of eye disorders).

Color Blindness

Color blindness, being a sex-linked hereditary condition, appears much more commonly among males than among females. There are three basic types of cones in the retina: red-sensitive, green-sensitive, and violet-sensitive. Any one or two or all three kinds of cones can be defective. Thus, there are seven possible types of color blindness. Total color blindness, in which everything is seen in black and white, is very rare. There is, so far, no prevention or cure for color blindness.

Conjunctivitis

Conjunctivitis is an infection or inflammation of the conjunctiva. The blood vessels become enlarged and visible, giving the eye a blood-shot appearance. If this condition becomes chronic or serious, it should be seen by an ophthalmologist, who can usually clear it up with prescribed medications.

Cataract

Cataract is the development of a cloudiness of the lens so that its normal transparency is reduced. The condition occurs mainly in older people, though it can even be present at birth. Severe cataract sometimes makes the pupil of the eye appear white. The only cure for cataract, to date, is to surgically remove the entire lens from the eye. The function of the lens is then assumed by eyeglasses. Since the eye then has reduced focusing mechanism, strong convex lenses are required. The person may use more than one pair of glasses to allow for close and distant vision.

Glaucoma

Glaucoma is the leading cause of blindness in the United States today. Glaucoma is an excessive pressure inside the eye. This excess pressure comes about when the aqueous humor is produced faster than it can escape through tiny channels to the outside. This pressure destroys the nerve fibers of the retina and pinches the blood vessels feeding the retina, so the cells of the retina are permanently destroyed. Blindness from glaucoma is permanent. Visual loss from glaucoma can be avoided by early diagnosis and proper treatment. But the disease is often overlooked in its early stages because it can steal sight so slowly and painlessly that many people do not realize they have the disorder until permanent damage is done.

Symptoms, when they appear, include headaches, loss of peripheral vision, and the illusion of halos around lights. The disease may develop slowly or very rapidly. The best way of diagnosing glaucoma is with the *tonometer,* an instrument which is pressed against the surface of the eye. The process is painless and harmless. After the age of thirty, this test should be part of every eye examination; after the age of forty, it should be performed at least every two years. Glaucoma is treated with drugs and surgery.

CARE OF THE EYES

We have seen how complex the eye is, and a few of the disorders which it may develop. Because the eyes are so essential to a person, they deserve the very best of care. Several professions deal with the eye. The *ophthalmologist* is a specially trained physician who can examine the eyes, prescribe lenses, give medication, and perform eye surgery. The *optometrist* is trained to examine the eyes and to prescribe appropriate lenses. He is not a physician and thus is neither trained nor permitted to perform eye surgery or prescribe any drugs for the eye. An *optician* is trained to grind lenses and mount them in frames to fit the face of the patient.

Contact Lenses

Contact lenses are small plastic lenses that ride on a thin layer of tears directly over the cornea and under the eyelids. Recent improvements in their construction have increased their popularity. Because they are practically invisible when worn, they are widely used by people in sports and entertainment and are preferred by many to the appearance of regular glasses.

Contact lenses cannot be used by everyone, however. Certain

types of visual defects can be corrected satisfactorily with contacts; others cannot. In addition contacts must be placed in the eye and removed daily, so they are easily lost. Some people find them too irritating to wear at all.

Contact lenses must be properly fitted to the eye by a qualified optometrist or ophthalmologist. It may take many adjustments before they can be worn comfortably; and this makes contact lenses *expensive*. There are those who advertise contact lenses at lower prices, but these individuals or organizations are often unwilling to make the adjustments necessary to make the lenses fit comfortably.

Sunglasses

Bright sunlight or glare can cause squinting and eyestrain and may contain harmful amounts of ultraviolet rays which can gradually damage the lens and retina. Sunglasses can reduce the discomfort of bright sun and screen out much of the ultraviolet rays. Another reason for wearing sunglasses is that driving vision during the early evening hours is considerably reduced if the eyes have been exposed to bright sunlight during the day.

The prices for sunglasses range from very inexpensive to very expensive. It is wise to pay a little more and get a good pair of sunglasses. First, they should be fairly dark. Second, they should be properly ground, as optical flaws will create eyestrain and headaches just as bad as produced by the bright sun. A person who normally wears prescription lenses should have sunglasses ground to his prescription or get the type that clip onto his glasses.

Sunglasses should *never* be worn for night driving or for looking directly at the sun.

Eye Exercises

Although much has been said in the past about the value of eye exercises, they are actually of value for only a few specific conditions. They are of no value in restoring diseased eyes or in changing the proportions of the eyeball. Eye exercises should be done only under the direction of an ophthalmologist.

Eye Irritations and Injuries

The eye is extremely vulnerable to injury by foreign objects and chemicals. Any object that becomes embedded in the eye, should be removed promptly and only by a physician. The use of any cosmetic, hairspray, or other product that causes eye irritation should be immediately discontinued. Many physicians warn against the use of eyewashes or drops. The natural flow of tears is the best means of cleaning the eye.

Every day, many eyes are blinded by flying objects or sharp

instruments or toys. Anyone working near a grinding wheel, or any other type of high speed machine which might throw off particles, should wear protective goggles. Children must be cautioned and closely supervised when they use scissors or play with any sharp-edged or pointed toy.

Proper Lighting

Despite what you may have heard, poor lighting will not "ruin" your eyes. But it can cause eyestrain, fatigue, headaches, and slow reading speed. Good reading light should be bright enough, evenly distributed, and free of glare. Try moving your reading lamp around until the best lighting position is found.

STRUCTURE OF THE EAR

The ear is a combination sense organ, functioning for both hearing and equilibrium. The ear may be divided into three parts: the *external ear,* the *middle ear,* and the *internal ear* (Figure 4.4).

The external ear consists of the visible part of the ear, called the *auricle* (ear) or the *pinna,* and the auditory canal. These structures pick up sound and carry it toward an opening in the skull. At the end of the auditory canal is the *tympanic membrane* (eardrum), which forms a barrier between the external ear and the middle ear.

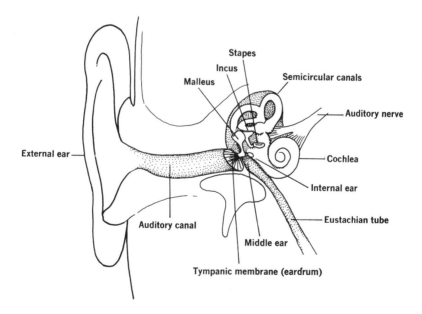

Figure 4.4 Structure of the ear.

The tympanic membrane is thin and stretched tight, so any sound waves that fall upon it cause it to vibrate.

The middle ear is an air space which contains three very small, very important bones called the *ossicles*. The bones transmit sound from the external ear to the internal ear, and are called the *malleus* (hammer), the *incus* (anvil), and the *stapes* (stirrup). The ossicles are united by true joints to form a lever system that amplifies the vibrations picked up from the eardrum and transfers them to a membrane, the *oval window*, on the internal ear.

The middle ear is connected to the throat by the *eustachian tube*. The purpose of this tube is to keep the air pressure in the middle ear equal to that on the outside. The end of the tube at the throat is closed most of the time by a valve which prevents one's own voice from sounding overly loud. The act of yawning or swallowing will temporarily open this valve so that air pressure can be equalized. Since the membrane that lines the throat extends up the eustachian tube into the middle ear, throat infections can easily follow this membrane into the middle ear.

The internal ear is the area of sensory reception for both hearing and equilibrium. Sound waves are converted to nerve impulses within the *cochlea*. These impulses travel to the brain by way of the *auditory nerve*. The organs of equilibrium include the three *semicircular canals* and, beneath them, the *saccule* and the *utricle*. The semicircular canals are sensitive to *changes* in position while the saccule and utricle inform a person of the actual position of his head.

DISORDERS OF THE EAR

The structures of the middle and internal ear are very delicate and can be easily damaged. Although these structures are embedded within the skull, they are exposed to the outside through both the auditory canal and the eustachian tube.

Few people who have not experienced deafness can realize the extent to which it interferes with a normal life. Total deafness means no conversation, no music, movies, television, or anything that includes sound. Fortunately, most hearing loss is partial, rather than total.

There are two basic types of hearing loss—conductive and perceptive. *Conductive deafness* occurs when there is some kind of interference with the passage of sound through the external and middle ear. It can be caused by obstruction of the auditory canal, damage to the eardrum, infection of the middle ear, or damage to the bones of the middle ear. If a person is aware of persistent "fullness" in the ears, ringing or buzzing sounds, or difficulty in hearing, he should promptly see a physician who specializes in ear disorders *(otologist)*.

Perceptive deafness is caused by damage to the sensory cells contained within the cochlea or damage to the auditory nerve. This form of deafness has many causes and is very common among older people. It may be the result of prolonged exposure to high sound levels, as in many factories or in highly amplified music. Many rock musicians have developed significant hearing losses while still in their teens. Other causes include diseases such as measles, mumps, or scarlet fever; injuries, such as blows to the head; or the effects of certain drugs.

CARE OF THE EARS

Several general suggestions can be made which can help a person avoid damage to his hearing.

1. Get prompt medical treatment for any ear or throat infection. Ear infection is one of the most common causes of hearing loss.

2. Blow your nose gently, if at all, so as not to force infections up into your ears.

3. Do not insert *any* object into the auditory canal. The eardrum or the delicate bones of the middle ear can be easily damaged.

4. Do not attempt to remove wax from the ears. If wax accumulation is a problem, ask a physician to show you how it may be safely removed.

5. Avoid overly loud sounds. Any sound that hurts is loud enough to cause permanent hearing loss, over a period of time. Stereophonic earphones turned-up loud can be very harmful. The sound of certain types of racing cars (such as fuel dragsters) can be damaging. Anyone who is repeatedly exposed to loud sounds should use ear plugs to avoid hearing loss. Hearing loss from loud sounds is so gradual that a person is unaware that it is occurring.

Hearing Aids

Many highly advanced models of hearing aids are made today which can greatly improve the hearing (in certain kinds of deafness) without being unsightly in appearance. No person of any age should hesitate to wear a hearing aid if his deafness is of a type that can benefit from its use. Before the purchase of a hearing aid, however, a physician should always be consulted for a complete ear examination to determine just what the problem is, and whether a hearing aid is the best solution.

SUMMARY

I. Structure of the Eye
   A. Cornea—"window on front."
   B. Crystalline lens adjusts for viewing at different distances.

C. Ciliary body controls the lens.
D. Suspensory ligament connects lens to ciliary body.
E. Aqueous humor fills space between cornea and lens.
F. Vitreous humor fills space behind lens.
G. Iris—colored part of eye; opening in center is pupil.
H. Sclera—outer layer of wall of eyeball, gives shape.
I. Choroid—membrane inside sclera; nourishes retina.
J. Retina—light-sensitive part of eye; features include:
   1. Rods are sensitive to black and white.
   2. Cones are sensitive to colors.
   3. Blind spot is over optic nerve, contains no rods or cones.
   4. Fovea—area of cones only; point of sharpest vision.
K. Conjunctiva is the membrane lining eyelids and covering front of sclera.
L. Tear glands produce tears to lubricate eyelids, clean eyes, and inhibit growth of bacteria.

II. Disorders of the Eye
   A. Focusing defects
      1. Myopia (nearsightedness)—light rays come to focal point in front of retina.
      2. Hyperopia (farsightedness)—light rays are focused at a point behind the retina.
      3. Astigmatism (mixed vision)—light rays do not all focus on same plane.
      4. Presbyopia—loss of elasticity of lens with age.
   B. Muscle imbalance causes eye to turn in, out, up, or down, a condition called strabismus.
   C. Color blindness—hereditary defect in cones.
   D. Conjunctivitis—infection or inflammation of conjunctiva.
   E. Cataract—cloudiness of lens.
   F. Glaucoma—excessive pressure inside eyeball.

III. Care of the Eyes
   A. Several professions
      1. Ophthalmologist is a specially trained physician; can examine eyes, prescribe lenses, give medication, and perform surgery.
      2. Optometrist is not a physician; can examine eyes and prescribe lenses, cannot perform surgery or prescribe drugs for eyes.
      3. Optician grinds lenses and mounts in frames.
   B. Contact lenses
      1. Ride on layer of tears over cornea.
      2. Must be properly fitted to eye.
   C. Sunglasses
      1. Reduce discomfort of bright sun and screen out harmful ultraviolet rays.

2. Should be of good quality.
D. Eye exercises are of value in only a few specific conditions.
E. Eye irritations and injuries
 1. Foreign objects should be removed promptly by physician.
 2. Many physicians warn against the use of eyewashes or drops.
F. Proper lighting prevents eye fatigue.

IV. Structure of the Ear
 A. External ear
  1. Auricle or pinna—visible part of ear.
  2. Auditory canal carries sound to middle ear.
  3. Tympanic membrane (eardrum) separates external ear and middle ear; picks up sound vibrations.
 B. Middle ear
  1. Ossicles (ear bones) transmit sound from eardrum to internal ear.
  2. Eustachian tube connects middle ear to throat; equalizes air pressure.
 C. Internal ear
  1. Cochlea converts sound waves to nerve impulses.
  2. Auditory nerve carries impulses to brain.
  3. Semicircular canals, saccule, and utricle are organs of equilibrium.

V. Disorders of the Ear
 A. Conductive deafness is caused by obstruction or damage in external or middle ear.
 B. Perceptive deafness is caused by damage to sensory cells in cochlea or to auditory nerve.

VI. Care of the Ears
 A. Get prompt treatment for any ear or throat infection.
 B. Blow nose gently, if at all.
 C. Do not insert any object into auditory canal.
 E. Do not attempt to remove wax from ears without instruction by physician.
 F. Avoid loud sounds.
 G. Hearing aids are of value in certain types of deafness.

QUESTIONS FOR REVIEW

1. Give the function of each of the following parts of the eye:
   a. cornea   f. choroid
   b. lens   g. retina
   c. ciliary body   h. rods
   d. iris   i. cones
   e. sclera   j. fovea
2. Contrast myopia and hyperopia.

3. What is astigmatism?
4. What is cataract?
5. What is the leading cause of blindness in the United States today?
6. Contrast the ophthalmologist and the optometrist.
7. Give the function of each of the following parts of the ear:
   a. auditory canal
   b. tympanic membrane
   c. ossicles
   d. eustachian tube
   e. cochlea
   f. semicircular canals
8. Name and contrast two basic types of deafness.
9. How can a sore throat threaten hearing?

Chapter 5
THE SKIN AND HAIR

The skin and hair are of particular interest to most people since they so greatly influence appearance. But, in addition, the skin plays many important roles in maintaining the health of the entire body.

STRUCTURE OF THE SKIN

The skin (Figure 5.1) consists of two layers. The outer, surface layer is the *epidermis*. This tough covering is less than one-sixteenth of an inch thick in most parts of the body. The outermost cells of the epidermis are dead and are constantly being worn away and replaced from beneath. The cells in the deeper part of the epidermis are alive and are continually dividing to replace the cells being lost from

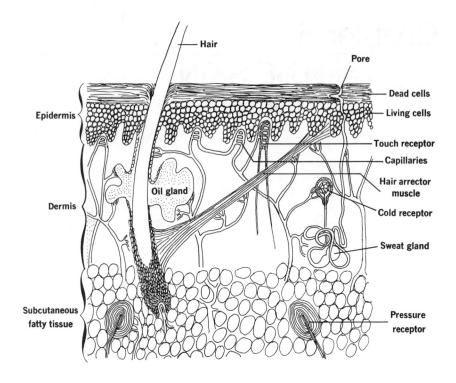

Figure 5.1 Cross section of human skin. Note hair follicle and sense receptors.

the surface. As they push outward, the new cells gradually receive less and less nutrition from the blood capillaries below them, until they eventually die. Their protein then converts from its living form to a tougher form called **keratin**.

Beneath the epidermis lies the *dermis*, sometimes called the true skin. The dermis is thicker than the epidermis, ranging from one-sixteenth up to one-quarter of an inch thick, depending upon the part of the body. The dermis contains many different kinds of tissues, including fibrous connective tissues which give it strength and toughness, an abundance of blood capillaries, nerves and sensory receptors, fat deposits, hair follicles, sweat glands, and oil glands.

Skin Color

The color of the skin is produced by several pigments. The most important of these is *melanin*, a yellow to black pigment which is produced in the living part of the epidermis by special cells called *melanocytes*. In heavily pigmented skin, melanin is present throughout the epidermis. The amount of melanin produced depends on heredity and sun exposure. Melanin is the tanning pigment; its production is stimulated by ultraviolet rays contained in sunlight. The main purpose of melanin is to absorb ultraviolet rays in order to prevent them from burning the deeper layers of skin. Albinos are people who lack one of the enzymes necessary to produce melanin.

The color of the skin is also influenced by the yellow pigment *carotene* and by the *hemoglobin* in the blood circulating through the dermis layer.

FUNCTIONS OF THE SKIN

The skin is much more than just an inert covering over the body. It is a *vital organ* which serves many functions necessary for life. If a large enough area of the skin is destroyed, death is certain to occur.

Protection

The healthy unbroken skin is a barrier against infection and mechanical or chemical injury to the delicate tissues beneath it. Its waterproof quality prevents the loss of the fluid that bathes the cells of the body. The sensory nerve endings in the skin give protection by alerting us to hazardous conditions such as objects which are dangerously sharp, hot, or cold.

Regulation of Body Temperature

It is very critical that the amount of heat lost from the body be in balance with the amount of heat produced. The average body temperature, as measured orally, is about 98.6° F. (Fahrenheit), but temperatures between 97° and 99° F. are considered normal. The rectal temperature runs about one degree above the oral temperature, while the underarm temperature runs about a degree less. The source of this heat is the activity of the body muscles and glands. As more heat is produced, the heat loss from the skin must increase accordingly.

Heat is lost from the skin in two ways. The first is by contact with the air or any other material which is cooler than the skin. Moving air, of course, carries away much more heat than still air. The second means of heat loss is through the evaporation of sweat from the skin. Since evaporating water absorbs much heat, there can be effective loss of body heat, even when the temperature of the air is greater than the body temperature.

The production of sweat goes on at all times, even in cold weather. We are generally only aware of our sweating when the rate of sweat production exceeds the rate of evaporation, as when the relative humidity of the air is high or the muscles are unusually active. When both the temperature and the humidity are high, it may be necessary to be less active in order to keep the body from overheating.

Overheated blood causes a reflex action whereby the small blood vessels in the dermis dilate (open further) so that more blood circulates near the surface. This helps increase the heat loss and causes a "flushed" appearance. When the air is cold and the body must conserve heat, these blood vessels close down.

Excretion

Not only does sweating serve to cool the body, but it is an important means of eliminating wastes as well. Sweat consists of about 99 percent water. The dissolved material in sweat is about half sodium chloride (table salt) with the remainder being mainly urea and other nitrogen-containing chemicals. Sweat contains traces of many very bad smelling chemicals which contribute to "body odor," especially when these chemicals are modified by the skin bacteria.

An inactive person in a perfectly air conditioned room will still sweat at least a pint in twenty-four hours. But, under extreme conditions of heat and work, sweat production may reach as much as three gallons in twenty-four hours. When sweating is profuse, the amount of salt lost from the body is greatly increased. If this salt is not replaced, serious physiological reactions may take place including nausea, vomiting, weakness, or even loss of consciousness. During hot weather an increase in the salt intake is often recommended, especially for the physically active person.

Production of Vitamin D

The skin of man contains large amounts of a substance (7-dehydrocholesterol) which can be converted into vitamin D by exposure of the skin to sunlight. Since most foods are low in vitamin D (unless it has been artifically added), the skin is an important source of this vitamin. The production of vitamin D is, of course, greater in the summer months. Since melanin filters out much of the ultraviolet radiation, darkly tanned or pigmented skin is less effective in the production of vitamin D.

Sensory Reception

The skin contains receptor structures for many different kinds of stimuli. There are separate receptors for heat, cold, touch, pressure, and pain. These receptors are of obvious importance in informing us of environmental factors or dangerous conditions, in allowing us to make delicate manipulations, and in sexual stimulation.

HAIR

Hair is distributed over almost the entire human body though, of course, its nature varies from area to area. Each hair arises from a deep pit in the skin, the *hair follicle*. The hair shaft consists of closely compacted, dead, keratinized cells, cemented together. At the base of the hair follicle is the onion-shaped *bulb* containing the *hair root*. The hair grows by cell division in the root. Each hair follicle goes through *hair cycles* in which a new hair develops, grows for a period of time, dies, and is replaced by another hair. Thus, it is normal for a certain amount of hair to be falling out at all times.

Each hair follicle has a tiny involuntary muscle which can cause the hair to "stand on end" in response to cold or fear. These muscles are the cause of the so-called "goose flesh" of the skin. Also attached to each hair follicle are one or more sebaceous (oil) glands which empty their fatty secretions onto the skin through the hair follicle.

NAILS

The structure of the nails is somewhat similar to that of the hair. The nails are densely compacted, dead, keratinized cells which grow from living cells at the base of the nails. The part of the nail that shows is called the *body;* the hidden part is the *root*. The crescent-shaped white area near the root is called the *lunule*. The fold of the skin that covers the base of the nail is the *cuticle*. The nails provide a firm resistance over the soft fingertips so that small objects can be picked up and held more easily. When a nail is torn off or falls off due to bruising of the tissue beneath it, a new one will grow in its place, provided the living cells at its base are left.

The nails require some special attention. To prevent hangnails and snagging of stockings, the toenails should be cut straight across, with the sharp corners rounded off. The fingernails are generally cut to conform to the tips of the fingers. The cuticles may be pushed back, if desired, but they should not be cut off because cutting them only stimulates their growth and results in thickened cuticles.

CARE OF THE SKIN

The care of the skin is of great importance because the condition of one's skin influences his appearance, his social relationships, and his physical and emotional health. The principles of good skin care are simple.

Cleanliness

The most important step in skin care is to keep it clean, head to toe. A shower or bath every day not only washes away dirt, bacteria, and excess oil, but it relaxes the muscles and stimulates the circulation of blood through the skin as well. The great majority of all skin problems can be prevented by simply keeping the skin clean. The socially aware person not only wants his skin to look good, but he wants it to smell good, too. When the oils of the skin and hair accumulate, the bacteria on the skin and scalp bring about changes in these oils which cause them to become rancid and odorous, much like stale butter. Stale sweat also has a very unpleasant odor, especially after the action of the skin bacteria. The bath soap should preferably be of a deodorant type, which means that it contains chemicals to inhibit the growth of bacteria. Such soaps not only help eliminate body odors, but are believed by many physicians to help keep the skin free of minor infections and blemishes.

A hot shower followed by rubbing the skin with a towel aids the circulation of blood through the skin and produces a feeling of well-being and renewal.

Because the greatest concentration of oil glands is on the face, many persons need to wash their face more than once a day, especially if acne is a problem.

Diet

Many cosmetics are claimed to be able to nourish the skin. Actually, the skin is nourished by the blood and the only way to make sure the necessary foods are available to the skin is to eat right. The nutritional requirements for the skin are the same as for the rest of the body—plenty of fruits, vegetables, meats, milk, and eggs.

Deodorants

Even with frequent bathing, it is usually necessary to use an underarm deodorant to help prevent unpleasant odor. There are a multitude of deodorants available, with many different formulas and methods of application. Some have a second function of reducing the activity of underarm sweat glands. Many people find that certain brands of deodorants cause irritations or rashes. If such a reaction is noted, other brands should be tried.

Cosmetics

Many false claims are made about the benefits of certain cosmetics. Small amounts of carefully applied cosmetics can highlight the appearance and emphasize certain features, but the constant heavy application of cosmetics does not improve the basic appearance of the skin, and often damages the complexion.

Cosmetics should be used around the eyes with extreme care, as they may irritate the eyes. Some girls have even had their eyelashes fall out from their eye makeup.

Some cosmetics contain substances that are harmful to the skin of certain persons, causing rashes or other reactions. It is a good idea to test a new cosmetic on your arm or body before putting it on your face. If your skin shows any reaction, you will know that you should not use that cosmetic.

Every new season brings its makeup fads. But the "natural" look is always in style. Healthy, naturally beautiful skin needs little makeup.

SPECIAL SKIN PROBLEMS
Infections

The skin, being in direct contact with the environment, is subject to many kinds of infections.

BOILS

A boil is an *abscess* (pus-filled cavity) caused by an infection with *Staphylococcus aureus*. *Staphylococci* are bacteria which cause many local infections. The pus formed is mainly a mixture of bacteria and white blood cells. In the past, many people thought that boils were an effort by the body to "purify" the blood, but that is not the case at all. When several boils occur close together, the mass is called a *carbuncle*.

A boil should *never* be pinched or squeezed. Pinching a boil could cause the bacteria to spread further into the surrounding tissues or even to enter the bloodstream and cause a very serious infection of the blood.

Small, simple boils should be protected from irritation by covering them with a sterile gauze compress. They usually come to a head in about a week, break open, discharge the pus core. The bandage should then be changed often until the healing is complete.

Large or painful boils, or those of the head, face, or neck, should be treated by a physician. He may have to lance the boil to relieve the pressure of the pus and he may also give an antibiotic to aid in the healing of the boil.

IMPETIGO

Impetigo is a surface infection of the skin. Small blister-like pustules appear on the skin and soon rupture to produce a "weeping" spot on the skin. A fluid oozes from this spot and dries to form a yellowish crust. The spots may burn or itch. The cause of impetigo is usually *Staphylococcus*, but it may also be *Streptococcus*.

Impetigo is highly contagious, either by direct or indirect contact with the discharges from the sores. The infection can easily be spread by anything that has been contaminated with the fluid discharge. Impetigo is common on the face, hands, arms, and legs. Its treatment generally requires the aid of a physician. It is important that the sores not be touched or scratched as this will further spread the infection.

ATHLETE'S FOOT AND RINGWORM

Athlete's foot and ringworm are general terms applied to infections by various fungi (molds). These fungi can infect the scalp, skin of any part of the body, and even the nails. The name ringworm comes from the tendency of the fungi to form sores that heal in the center while spreading outward to form a ring. It is not caused by worms. Fungus infections are among the most difficult to cure, and sometimes last for years despite medical treatment. These fungi spread by producing millions of microscopic spores.

Ringworm of the scalp can be caught from infected persons or animals, including dogs, cats, cattle, and horses. It can be caught indirectly from contaminated theater seats, hats, barbers' tools, borrowed combs, and similar sources. Untreated cases can result in spots of temporary or permanent baldness. A physician should be consulted promptly for its treatment.

Ringworm of the body can be caught from infected people or animals, or indirectly from contaminated clothing, shower stalls, benches, and similar articles. It can sometimes be remedied by a thorough washing with soap and water and home treatment with ointments, but treatment by a physician is often necessary.

Ringworm of the foot is *athlete's foot*. There is a scaling or cracking of the skin, especially between the toes. Characteristic blisters, filled with a thin, watery fluid, form; following which the skin cracks and peels off. It may itch, burn, or even bleed.

The spores of athlete's foot are likely to be present anywhere that people walk barefooted. In many places, such as swimming pools, it is just impossible to avoid picking up these spores on the feet. Since the fungus cannot be avoided, it is important to deny it the conditions it needs to thrive. First, the feet must be kept *clean*—washed often and thoroughly with soap and water. Second, the feet must be kept *dry*. This requires a clean, dry pair of socks every day. The same pair of shoes should never be worn two days in a row, so they can dry out between wearings. Any type of shoe or sock that interferes with the evaporation of sweat from the feet should be avoided. Leather soles are generally preferable to rubber ones for this purpose. Dry feet carefully after showers. Sprinkling a foot powder on the toes, in the socks, and in the shoes is often helpful. Treatment by a physician may be necessary.

Ringworm of the nails can affect the hands or feet. The nails gradually become thickened, rough, discolored, and brittle. The disease is more common in older persons (males more than females) and it is not highly contagious. A physician should be consulted for its treatment.

SCABIES

Scabies is the infestation of the skin by microscopic crab-shaped mites which are just barely visible to the naked eye. These mites live in tiny burrows they make in the skin. They often occur around the finger webs, wrists, elbows, thighs, abdomen, and sex organs. Symptoms include a rash and intense itching. (The disease is sometimes called the seven-year itch). Scratching should be avoided as it only acts to spread the infection and allow entry of secondary infection by bacteria.

Scabies can spread by direct contact (it is often caught during sexual intercourse with an infected person) or indirect contact such as through clothing or bedclothes. A physician should be consulted for its treatment.

LICE

There are three types of lice which commonly infect humans. They are all small insects, between one-eighth and one-sixteenth inch long. The first of these is the *head louse* which most commonly lives on the head, but can live on other hairy parts of the body. In severe infestations, the hair may become matted with louse eggs (called *nits*), lice, and a crust which is the dried up fluid that has oozed from the louse bites on the scalp. Head lice are easily caught by direct contact with infested persons or by stray, loose hairs, combs, and barbers' tools. They can be easily eradicated under the direction of a physician.

The *body louse* is sometimes called the "*cootie*." It is very closely related to the head louse and very similar in appearance, but its habits are different. They may live on either hairy or smooth parts of the body. These body lice spend much of their time in the clothing and perfer to lay their eggs in the seams of clothing, rather than on hairs. Louse infestation is mainly caught by close contact with infected persons or their clothing. The feeding of these lice can cause a general tired feeling and skin rash. Body lice are known to spread several serious diseases.

The third type of louse is the *pubic louse* or *crab louse*, often just called "crabs." These lice live mainly in the pubic hair, but occasionally in other body hair. They grasp the hairs with their large claws, and may spend days attached in the same place, with their sucking beaks inserted into the skin. This feeding causes an intense itching and often some bleeding. Crab lice are transmitted from person to person by sexual intercourse or other direct or indirect contact. They are best treated under the direction of a physician.

Acne, Blackheads, and Pimples

Acne can be one of the most troublesome problems of youth. The skin appears greasy; it is marked by blackheads, red spots, and pus-filled pimples. The person with acne may feel self-conscious (it probably looks worse to him than it does to other people). But severe cases of acne can leave scars on the skin that may last for years.

The cause of acne seems to be rather complex, usually involving overproduction by the oil glands of the face, back, and chest. The openings of the oil glands get clogged with oil and dirt, become enlarged, and become infected with bacteria. These enlarged oil glands are called *blackheads* if they are tipped with dirt and *whiteheads* or *pimples* if they are infected and filled with pus. Sexual activity, or the lack of it, has nothing at all to do with acne.

Many cases of acne can be cleared up with proper home care. There are several simple steps to this care.

1. *Proper diet is important.* The best diet for fighting acne includes plenty of fruits and vegetables and lean meats. Sweets, starchy, and fatty foods should be avoided.

2. *Cleanliness.* The face and any other affected parts should be thoroughly washed three or four times a day with plenty of hot water and soap. A deodorant or disinfectant type soap is of value for inhibiting the infecting bacteria, so long as it does not irritate the skin. Gently rub the lather into the skin for several minutes to remove the excess oil. Rinse first with hot water to open the pores, then rinse again with cold water to close the pores. Over a period of time this will remove the blackheads and other types of plugged pores.

3. *Don't squeeze pimples or blackheads.* This often spreads the infection, enlarges the pimples, and may even force bacteria into the blood. A pimple which has been squeezed and picked may leave a permanent scar. The best way to clear clogged pores is described in the above paragraph.

4. *Use cosmetics very sparingly.* Use no oils or creams on the affected parts. In acne there is already too much oil present. Many types of cosmetics also contain fine particles which may clog the pores.

If acne does not clear up with two or three weeks of the home care just described, *consult a physician promptly*. If treatment is delayed too long, there may be some permanent scarring of the skin. While there are techniques for removing much of the scar tissue today, it is far better to prevent the scarring in the first place.

Suntans and Sunburns

Many lightly pigmented people feel that they look better with a suntan. Some people are able to tan beautifully; others have skin types that just do not tan. But even the person who tans well must be careful to avoid sunburn at the beginning of the summer. The key is to start with short periods of exposure and gradually work up to longer times. A very white person may be starting to burn after as little as fifteen minutes in the sun. But this will stimulate the production of enough pigment that the next day's exposure can be increased to perhaps thirty minutes. After a few days, exposures of several hours may be possible. We hate to think how many millions of vacations have been spoiled by a bad sunburn on the first day.

Some brands of suntan creams and lotions are effective enough to somewhat extend the time one can stay in the sun. The action of these preparations is to screen out the ultraviolet rays, which are responsible for both tanning and burning. Despite the advertisements, there is no way to screen out the "burning rays" of the sun while admitting the "tanning rays," since they are one and the same.

Those who like to tan very deeply every year should be aware of several possible effects of extreme sun exposure. One is the development of premature wrinkles caused by the aging effects of ultraviolet rays on the skin. The skin may actually be made to look many years older than its true age. Another possible result is the development of skin cancer. Skin cancer is most common on the constantly-exposed parts of the skin, such as the face, neck, and ears. It is more common among people who suntan with difficulty than among those who tan easily.

Warts, Moles, and Birthmarks

Warts are caused by a virus, not by handling toads. They often disappear spontaneously, without any treatment, which is why so many home remedies for warts are thought to have worked. There is actually no way to remove warts at home that is both safe and effective. If a person wants a wart removed, he should consult a physician, who can use one of several very simple, safe, and effective methods of wart removal.

Moles are pigmented growths in the lower layer of the skin. Certain types of moles occasionally become cancerous. Any mole that suddenly changes color, or begins to grow, should be checked by a physician.

Birthmarks are discolored areas of skin, present at birth. Red birthmarks are usually due to enlarged blood vessels just under the skin; brown ones are due to excess melanin. The cause of birthmarks is unkown. They have nothing to do with any thoughts or actions of the mother during pregnancy. Certain kinds of birthmarks can be removed by physicians, but a person should never try to remove one himself.

Allergies

One of the most common places for allergies to be exhibited is the skin. These allergies often take the form of hives, eczema, and various rashes. Allergies are commonly developed against such varied substances as pollens, feathers, animal hair, dusts, cosmetics, and foods of all types. An allergy is a reaction to a substance which may be harmless to other people. The substance which produces the reaction is called an *allergen,* which means that it acts as an *antigen,* stimulating the production of *antibodies.* Antigens are, in turn, destroyed by antibodies. (Antigens and antibodies are further discussed in Chapter 6.) The symptoms of allergies are largely due to *histamine,* a chemical released from nearby tissues in response to antigen-antibody reactions.

The cause of an allergy can often be determined through a series of skin-patch tests, in which small amounts of various antigens, such as foods and cosmetics, are scratched into the skin by a physician. In these tests allergic reactions are indicated by reddening of the skin.

CARE OF THE HAIR

Clean, well-groomed hair adds to anyone's appearance. Dirty, neglected hair can cancel the effects of all other efforts made to look and dress well.

Shampooing

The hair and scalp need to be washed frequently. The oil glands of the scalp are very active in most persons. The oil that they produce accumulates on the scalp and hair, becoming rancid and smelling quite bad, and causing the hair to look stringy as well. This oil also works down onto the forehead and contributes to acne.

The exact frequency with which the hair should be washed depends on the amount of oil produced and the exposure to dirt. Many hair experts today suggest weekly washing as a minimum, with more frequent washing recommended for those with more oily scalp. Shampoos vary in their ability to remove oil. The person with oily scalp needs a shampoo which removes much oil; the person with dry scalp should select a shampoo intended especially for that condition.

Dandruff

There is no single cause, nor sure cure, for dandruff. Since the scalp loses dead cells from its surface just as does the rest of the skin, it is reasonable to expect at least some flaking off of epidermis from the scalp. This is a perfectly normal and unavoidable condition. But excess dandruff can be a real problem, usually falling into one of three basic types. The first of these is associated with a very dry scalp. It can often be helped by using a dry-scalp shampoo or hair-oil preparation. The second type results from too much oil, causing the dead skin cells to cling together in clusters forming dandruff flakes. This type of dandruff may be helped by more frequent shampooing. A third type of dandruff includes thick oily crust or scales over the scalp. This may indicate an infection and should be checked by a physician.

Baldness

While in their late teens and early twenties, many boys (and a few girls) notice that their hair is beginning to thin and that their hairline is pushing back. Anyone who is starting to lose hair faster than it is growing back, should consult with a dermatologist (physician specializing in skin disorders) to make sure that the condition is not due to an infection. Hair falling in irregular patches is especially likely to be due to infection. Baldness from infection is often preventable. Hereditary pattern baldness (the most common type) is still incurable.

SUMMARY
  I. Structure of the Skin
     A. Epidermis—outer layer; thin; cells at surface dead.
     B. Dermis—inner layer; thicker; many kinds of tissues.
     C. Color produced by pigments melanin, carotene, and hemoglobin.
 II. Functions of the Skin
     A. Protection.
     B. Regulation of body temperature.
     C. Excretion of wastes.
     D. Production of vitamin D.
     E. Sensory reception.
III. Hair
     A. Hair shaft consists of dead cells, cemented together.
     B. Hairs are periodically replaced.
 IV. Nails
     A. Structure is similar to hair.
     B. Grow from living cells at base.
  V. Care of the Skin
     A. Cleanliness is most important step.
     B. Good diet is necessary for healthy skin.
     C. Underarm deodorants are useful for odor prevention.
     D. Cosmetics should be used sparingly and with care.

72 THE HUMAN BODY

VI. Special Skin Problems
   A. Infections
      1. Boils—abscesses caused by *Staphylococcus aureus;* must never be pinched or squeezed.
      2. Impetigo—highly contagious surface infection of skin by *Staphylococcus* or *Streptococcus*.
      3. Athlete's foot and ringworm—fungus infections of the skin.
      4. Scabies—infestation of the skin by microscopic mites.
      5. Lice
         a. Head lice.
         b. Body lice.
         c. Pubic lice (crabs).
   B. Acne, Blackheads, and Pimples
      1. Cause is complex.
      2. Treatment includes the following:
         a. Follow a proper diet.
         b. Maintain high standards of cleanliness.
         c. Don't squeeze pimples or blackheads.
         d. Use cosmetics very sparingly.
         e. Consult a physician promptly if acne persists.
   C. Suntans and Sunburns
      1. Both are caused by ultraviolet rays.
      2. Excess sun can age skin prematurely or cause skin cancer.
   D. Warts, Moles, and Birthmarks
      1. Warts are caused by a virus.
      2. Moles should be watched for possible cancerous changes in size or color.
      3. The cause of birthmarks is unknown.
   E. Allergies—result of antigen-antibody reaction.
VII. Care of the Hair.
   A. Wash hair at least once a week.
   B. There is no single cause nor cure for dandruff.
   C. Baldness should be checked by a dermatologist (physician) to determine cause.
      1. Infection—often curable.
      2. Hereditary pattern baldness—not yet curable.

QUESTIONS FOR REVIEW

1. Compare the dermis and epidermis.
2. List five functions of the skin.
3. What are "hair cycles"?
4. How is athlete's foot best prevented?
5. Discuss the treatment of acne.

6. What are some possible effects of excess sun exposure?
7. What are some different causes of dandruff? How is each best remedied?
8. Is hereditary pattern baldness curable?

Chapter 6
THE CIRCULATORY SYSTEM

The human body consists of millions of cells. To function properly, or even to survive, these cells must constantly be provided with blood. Blood brings oxygen and foods to the cell, carries away the cell's wastes, and must be kept in ceaseless motion. The circulatory system of the body is designed to deliver blood to and from the capillaries where the vital exchange of materials occurs. Stray body fluids outside of the blood vessels must be picked up and returned to the major blood vessels. The circulatory system includes the heart, the arterial vessels, the capillaries, the venous vessels, the lymphatics, as well as the blood itself.

THE HEART

About the size of a person's fist, the heart is located just beneath the breastbone in the center of a person's chest (Figure 6.1).

THE CIRCULATORY SYSTEM 75

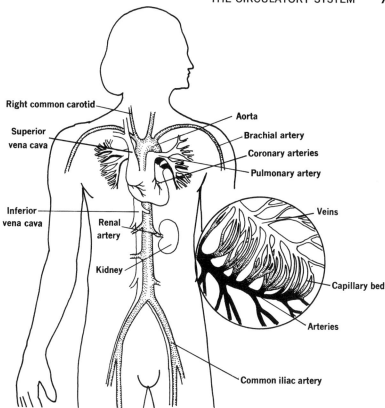

Figure 6.1 Human circulatory system showing principal arteries and veins. Insert, the relationship between arteries, capillaries, and veins.

Shaped like a cone, its tip, or apex, is directed downward and to the left. The lungs surround the heart on each side. The main mass of the heart wall is the *myocardium,* or the cardiac muscle. The heart is lined on the inside with a thin tissue called *endocardium.* The *pericardial sac* encloses it on the outside.

Heart Chambers

The heart contains four chambers, or cavities (Figure 6.2). The *right atrium* and *left atrium* are thin-walled chambers located in the upper part of the heart; the *right ventricle* and *left ventricle* are heavy-walled chambers below. The two atria are separated by a thin, muscular *interatrial septum;* the two ventricles are separated by a thicker, muscular *interventricular septum.* Two openings connect the chambers on either side of the heart and each opening is guarded by a flaplike valve. These valves permit blood to flow from the atria into the ventricles, but not in reverse. The right valve has three *cusps* and is called the *tricuspid valve;* the left valve is called the *mitral valve.* The right ventricle opens into the *pulmonary artery,* which carries

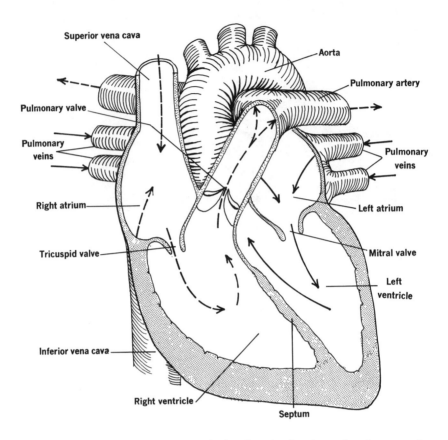

Figure 6.2 Internal structure of the heart showing chambers and route the blood follows.

blood to the lungs, and the left ventricle opens into the *aorta*, which carries blood to the rest of the body. Valves are located at the base of both of these blood vessels to prevent the reverse flow of blood. The *pulmonary valve* is at the base of the pulmonary artery, and the *aortic valve* is at the base of the aorta.

THE GENERAL CIRCULATORY SCHEME

Blood enters the left atrium of the heart from the lungs. It then flows into the left ventricle which pumps it out into the aorta for distribution throughout the body. The *arterial branches* of the aorta eventually become smaller and smaller until the blood reaches tiny *capillaries,* where oxygen and nutrients leave the blood and enter the intercellular spaces (Figure 6.1).

After passing through the capillaries, the blood is collected in the venules. The venules unite to form small veins which in turn combine to form larger veins that return the blood to the right atrium of the heart.

THE CORONARY VESSELS

The heart muscle itself is supplied by the coronary vessels. The coronary artery arises from the aorta as it leaves the heart and branches to feed both sides of the heart. The coronary veins collect the blood and return it to the right atrium.

PHYSIOLOGY OF THE HEART

The Heartbeat

When the body is at rest, more than ten pints of blood pass through each of the heart's chambers each minute. During strenuous activity, the capacity of the heart increases five to ten times. This organ begins its activities months before birth and never ceases until death.

The heart muscle operates in a manner completely different from that of any other muscle of the body. Although the heart is under the control of the central nervous system, its beat originates independently inside the heart. The beat starts in the right atrium at a spot called the *sinoatrial node* (SA *node*), also called the "pacemaker." From the right atrium the impulse travels over the entire atrial muscle, causing both right and left atrium to contract simultaneously. Blood in the right atrium is driven through the tricuspid valve into the right ventricle at the same instant that blood from the left atrium is driven through the mitral valve into the left ventricle.

From the SA node, a wave action is passed to a second specialized area, the *atrioventricular node* (AV *node*), located in the septum between the right atrium and ventricle. From this spot the wave action is transmitted to the *bundle of His*, a band of specialized cells which pass down the interventricular septum and divide into two main branches, called the *Purkinje fibers*. These fibers spread throughout the ventricular walls, causing both right and left ventricles to contract simultaneously. Blood in the right ventricle is driven through the pulmonary valve into the pulmonary artery at the same instant that blood from the left ventricle is driven through the aortic valve into the aorta.

To give the muscle some measure of rest, there is a rhythmic alternation of the contraction phase (*systole*) with a resting phase (*diastole*). The time needed for contraction and relaxation varies. At the usual rate of approximately seventy beats per minute in the adult, the diastole is twice as long as the systole. The systolic phase results in the familiar "lubb-dupp" sounds of the heart. The "lubb" sounds are caused mostly by the closure of the tricuspid and mitral valves; the "dupp" sounds by the closure of the pulmonary and aortic valves. If a valve fails to close properly, blood moves back through the valve, causing a "swishing" sound, called a *murmur*.

Pulse and Blood Pressure

With each beat of the heart, blood is forced out of each ventricle. Blood flows into the aorta with great force. The rhythmic contraction of the heart makes this discharge of blood rapid, spurtlike, and intermittent. During the interval between each beat, the heart collects as much blood from the body as it discharges. With each beat of the left ventricle, a wave of pressure starts at the heart and travels along the arteries. This wave is called the *pulse*. The pulse can be felt on any arteries that are close to the surface of the body, such as on the wrist, the sides of the throat, or the temple. The pulse is the result of the pressure of the blood on the walls of the arteries, or the *blood pressure*. This pressure is highest in the aorta; it gradually decreases as it travels through the arteries, capillaries, veins, and into the right atrium. Although the pressure is greatest after the left ventricle contracts, there is always some pressure in the arteries. The blood pressure at the moment of contraction is the *systolic pressure*; it should normally be sufficient to displace about 120 mm. (millimeters) of mercury in a glass tube. The blood pressure at the moment of relaxation of the heart is the *diastolic pressure*; it should normally displace about 80 mm. of mercury. Thus the average blood pressure in a normal young adult should be about 120/80 ("one-twenty over eighty").

The usual instrument for determining arterial blood pressure is called a *sphygmomanometer*. It consists of a rubber cuff wrapped around the arm and connected by a tube to a scaled mercury column. The cuff is inflated so that it collapses the brachial artery in the arm. By listening to the flow of blood through the artery with a stethoscope as the air in the cuff is let out, a physician can determine systolic and diastolic pressure.

The Lymphatic System

The food and oxygen leave the blood in the capillaries and enter the tissue fluid. This tissue fluid, which bathes all cells of the body and which acts as a connecting link between the blood and the cells, is known as *lymph*. It consists of certain fluid portions of the blood and white blood cells. Lymph accumulates faster than it is able to move back into the blood capillaries. As as result, it must be continually drained from the tissue spaces through a system of vessels called the *lymphatic system* (Figure 6.3). Throughout the body, in all the tissue spaces, there are thin-walled, delicate lymphatic capillaries.

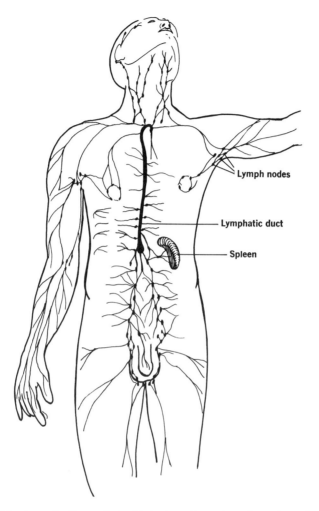

Figure 6.3 Lymphatic system. Note lymph nodes along lymph vessels which filter lymph.

These blind (closed), collecting ducts converge into larger lymphatic vessels which ultimately flow into veins near the base of the neck.

The lymphatic vessels have valves inside them to prevent backflow. Since the lymphatic system has no pump (heart) to push the lymph along, exercise and changes in body position help to maintain the flow of the lymph.

Along the larger lymph vessels are located a series of filters called *lymph nodes* (*lymph glands*). These nodes filter the lymph and produce white blood cells called *lymphocytes* which are added to the lymph. Lymph nodes are abundant in the groin, armpits, within the torso, and along the sides of the neck. Infection in the body can cause swelling and soreness at these locations.

FETAL CIRCULATION

During fetal life the blood of the unborn child travels through the umbilical arteries (in the umbilical cord) to the mother to be oxygenated. It returns to the child from the mother by way of the umbilical vein and moves into the right atrium. Since this blood is well oxygenated (the lungs of the fetus are not yet breathing), much of the blood entering the right atrium bypasses the lungs.

The fetus is normally provided with two short circuits—an opening in the interatrial septum (the *foramen ovale*) and a connection between the pulmonary artery and the aorta (*the ductus arteriosus*). At birth the child's connection to the mother is broken and he begins using his own lungs. Shortly after birth, both of these bypasses close. A child suffering from incomplete closure is known as a "blue baby".

BLOOD

Blood is an unusual liquid tissue that is kept in constant movement through the vessels of the body by the pumping action of the heart. It supplies cells deep within the body with the necessary materials to survive and function.

Within the average-sized individual there is about five quarts of blood. A healthy adult can tolerate the loss of one pint of blood, but is in danger of shock or death with the loss of three or more pints.

Blood consists of a straw-colored fluid, or *plasma*, within which are suspended *corpuscles*—red blood cells (*erythrocytes*), white blood cells (*leukocytes*), and *platelets* (Figure 6.4).

Plasma

Slightly over one-half of whole blood consists of plasma. Most of this plasma is a water solution containing many proteins, inorganic salts, vitamins, hormones, blood clotting factors, and glucose. Plasma also delivers oxygen and nutrients to the cells and carries away carbon dioxide and other wastes. When the blood clotting factors have been removed from plasma (by allowing it to clot) it is known as *serum*.

Red Blood Cells

The red blood cells (erythrocytes) are tiny disc-shaped bodies. Formed in the bone marrow, they usually live no longer than 120 days. Their main purpose is to carry oxygen from the lungs to the tissues. This is made possible by the pigment *hemoglobin*, an iron-rich protein which gives the cells their red color. The more oxygen present, the brighter red their color.

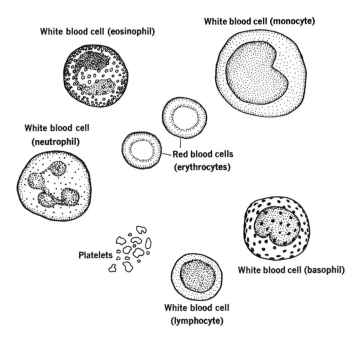

Figure 6.4 Blood cells.

White Blood Cells

The white blood cells (leukocytes) consist of various types of colorless cells which are formed in the bone marrow and lymph nodes. Although generally spherical, they have the ability to change shape. Such flexiblility allows some of them to leave the capillaries and move through the tissue spaces (which red blood cells cannot do). Some types contain granules (*neutrophils, eosinophils,* and *basophils*) while others contain none (*lymphocytes and monocytes*). Monocytes and neutrophils play an active part in protecting the body against infection, and in destroying bacteria and dead tissue by the process of *phagocytosis.* In this process the blood cells engulf, kill, and digest bacteria. Sometimes the white blood cells are affected by the poisons given off by the pathogens and die. They then become a part of the accumulation of material known as *pus.* Other types of white blood cells are involved in the manufacture of certain antibodies and in levels of immunity.

Platelets

Derived from large bone marrow cells which fragment, *platelets* or *thrombocytes* are essential to blood clotting, or *coagulation.*

When they come in contact with a blood vessel injury, they adhere to the damaged area. They then disintegrate and release a chemical which reacts with the protein in the plasma. *Fibrinogen,* a soluble plasma protein, with the aid of certain other blood components, changes into a solid mass called *fibrin,* which develops the clot.

Blood Count

The normal red blood cell count is 4.5 to 5.5 million per cubic millimeter of blood, as compared with a white blood cell count of 6000 to 8000 per cubic millimeter. This is a ratio of red cells to white cells of about 700–800 to 1. Blood counts are generally made with an apparatus called a *hemocytometer.*

Since the number of red and white cells is altered rather quickly in response to certain physical conditions, a blood count is an important diagnostic tool.

BLOOD GROUPS AND TRANSFUSIONS

If, for one of a number of reasons, the amount of blood in the body is too severely reduced, a person may need additional blood in the form of a *transfusion.*

Transfusions were often disastrous until it was discovered that red blood cells contained specific proteins (called *agglutinogens*).

ABO System

In the most widely used blood typing system (the *ABO System*), two main antigen types are identified, A and B. A person with antigen A in the red blood cells is classified Group A; one with antigen B, Group B; one with both antigens A and B, Group AB; and one with neither of these antigens, Group O. The percentage of each group in a population varies with the race. About 43% belong to Group O; 40%, Group A; 13%, Group B; and 4%, Group AB. A person's plasma contains other proteins called *antibodies* which complement the red cell antigens. There are two such antibodies, anti-a and anti-b. The plasma of a person of Group B contains anti-a; a Group A person has anti-b; a Group AB person, *neither* anti-a nor anti-b; a Group O person, *both* anti-a and anti-b.

The significance of the antibodies becomes apparent if Group A blood is crossed (mixed) with blood containing anti-a. The anti-a will cause the red cells to gather together in clumps (*agglutinate*) and seriously impede circulation. If too extensive, death can result.

In transfusions it must be firmly established that the *recipient's plasma* will not destroy the *donor's red cells.* It is always best to give blood of the recipient's own Group. [In an emergency it may be necessary to give (donate) blood to a recipient of Group O, A, B, or AB. Group O persons can give blood to all these Groups and are called *universal donors.*] Group AB persons can receive blood from donors of Groups O, A, B, and AB, and are called *universal acceptors.* Prior to transfusion, small volumes of both the donor's and recipient's blood should be *cross-matched* (mixed on a glass slide) to determine compatibility. Absence of agglutination indicates the compatibility of the two blood specimens.

Rh Factor

Among a number of other blood proteins is one called the *Rh (Rhesus) factor* which is found in the blood of about five out of six Americans. Those having it are called Rh positive (+); those lacking it, Rh negative (−). If an Rh negative person receives Rh positive blood (through transfusion or because an Rh negative mother is pregnant with an Rh positive child and some blood seeps across the placenta) the foreign Rh positive blood stimulates the production of an antibody, *anti-Rh*. This antibody can travel across the placenta. Produced in the body of the Rh negative woman, this antibody can enter the blood stream of the Rh positive fetus (back across the placenta) and cause the destruction of the red blood cells.

This destruction does not take place the first time Rh positive blood gets into the blood stream of the Rh negative woman, such as on the first transfusion or with the first pregnancy. The destructive action of the antibody may occur upon a second mixing or thereafter.

The Rh factor may be a problem in pregnancy only where the father is Rh positive and the mother, who is Rh negative, is carrying an Rh positive fetus. The anti-Rh antibodies produced by the mother's body destroy the fetus' red blood cells causing a severe type of anemia (inability of the blood to carry oxygen) and jaundice. To compensate, the bone marrow releases erythroblasts (immature red blood cells). In severe cases the infant may be born dead or may die shortly after birth. He may survive but be neurologically damaged.

Today a physician can make tests indicating a dangerous anti-Rh level during pregnancy. Damage can often be prevented by early induction of labor or an exchange blood transfusion to the infant immediately after birth or even before birth. An immune globulin product (RhoGAM) should be given to every Rh negative woman who gives birth to an Rh positive infant within seventy-two hours of such a delivery. This preparation of anti-Rh antibodies in effect "cancels the Rh effect of that pregnancy, thus protecting the *next* Rh positive fetus the woman might carry. Of course, this treatment must start with the *first pregnancy.*

SUMMARY
  I. The Heart
     A. Muscle is myocardium; inner lining is endocardium; outer covering is pericardial sac.
     B. Four chambers
        1. Upper two are atria.
        2. Lower two are ventricles.
     C. Walls between chambers are septa.
     D. Four valves prevent reverse flow of blood.
        1. Tricuspid valve is between right atrium and right ventricle.
        2. Mitral valve is between left atrium and left ventricle.
        3. Pulmonary valve is at base of pulmonary artery.
        4. Aortic valve is at base of aorta.
     E. Right side of heart pumps blood from body to lungs; left side of heart pumps blood returned from lungs back to body.
     F. Coronary vessels supply blood to heart muscle.
     G. Heartbeat originates in right atrium at sinoatrial node, the "pacemaker."
     H. Beat is passed to ventricles by atrioventricular node in septum between ventricles.
     I. Contraction phase of heartbeat is called systole; resting phase is diastole.
     J. Each heartbeat starts a wave of pressure in the arteries called the pulse; the pressure of the blood itself on the walls of the arteries is the blood pressure. Blood pressure is measured with a sphygmomanometer.
 II. Lymphatic System
     A. Lymph (tissue fluid) consists of certain fluid portions of blood and white blood cells.
     B. One function of lymphatic system is to drain off excess lymph.
     C. Exercise and changes in body position assist movement of lymph through lymphatic system; it has no pump.
     D. Lymph nodes (lymph glands) filter lymph and produce lymphocytes.
III. Fetal Circulation
     A. Much of the blood of fetus bypasses lungs, being oxygenated by placenta.
     B. Foramen ovale allows blood to flow from right atrium to left atrium; ductus arteriosus carries blood from pulmonary artery to aorta.
 IV. Blood
     A. Plasma—fluid portion; when clotting factors are removed is called serum.

B. Red blood cells
   1. Formed in bone marrow.
   2. Contain oxygen-carrying pigment, hemoglobin.
C. White blood cells—several types.
   1. Some produce antibodies.
   2. Some engulf and digest bacteria (phagocytosis).
D. Platelets, upon tissue injury, release a substance that initiates blood clotting.
E. Blood count
   1. Normal red cell count is 4.5 to 5.5 million per cubic millimeter of blood.
   2. Normal white cell count is 6000 to 8000 per cubic millimeter.

V. Blood Groups and Transfusions
   A. ABO system
      1. About 43% are type O; 40% A; 13% B; and 4% AB.
      2. Mismatching of blood transfusions results in agglutination (clumping) of red blood cells.
   B. Rh factor
      1. Present in about 5 out of 6 Americans, Rh positive (+); absent in 1 out of 6, Rh negative (—).
      2. Must not transfuse positive blood to negative person.
      3. May complicate pregnancies where Rh negative woman carries Rh positive fetus—now largely preventable with blood derivative called RhoGAM, injected into woman after each such pregnancy.

QUESTIONS FOR REVIEW
1. Name the four chambers of the heart.
2. What is the function of the heart valves?
3. How is the heart muscle itself nourished and supplied with oxygen?
4. Where does the heartbeat originate?
5. What does the blood pressure reading of 120/80 mean?
6. Trace the flow of blood from the right foot to the heart and back to the right foot.
7. What are the functions of the lymphatic system?
8. What is lymph?
9. How does exercise aid the lymphatic system?
10. What happens if blood, mismatched for ABO grouping, is transfused?

Chapter 7
THE RESPIRATORY SYSTEM

The energy for all the activities of the human body is derived from chemical reactions termed *biological oxidation*. Essentially this process takes place by the transfer of hydrogen atoms from chemicals derived from energy producing foods (carbohydrates, proteins, and fats). Through a series of compounds, hydrogen is accepted and donated until the ultimate hydrogen acceptor, *oxygen*, is bonded to the hydrogen, producing water.

Also, *carbon dioxide* is being removed (decarboxylation) from the food during this energy producing process. The carbon dioxide must be removed from the cell and body fluids as it is produced, for it reacts with water to form carbonic acid which is toxic in large amounts.

Since only small amounts of oxygen can be stored (as *oxyhe-*

moglobin in blood and *oxymyoglobin* in muscle), the energy producing process and the life of a human are dependent upon an uninterrupted supply of oxygen to each cell in the body. The respiratory system is responsible for supplying oxygen and removing carbon dioxide.

RESPIRATION

The term respiration has three different meanings. *Cellular respiration* refers to the processes (biological oxidation) by which cells utilize oxygen, produce carbon dioxide, and convert food into biological energy.

As shown by Figure 7.1, when there is an exchange of oxygen or carbon dioxide between body cells, body fluids, and the capillaries of the blood stream (see Chapter 6), the term *internal respiration* is used.

For the exchange of these gases between the blood stream and the air of the external environment the term *external respiration* is applied (Figure 7.1).

Between internal and external respiration the gases are transported by the circulatory system. Only external respiration is a function of the respiratory system.

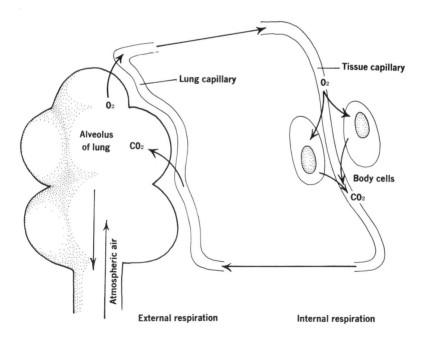

Figure 7.1 Diagram illustrating the exchanges of gases occurring in external and internal respiration.

Diffusion of Gases

The movement of oxygen and carbon dioxide across cell membranes is a physical process called *diffusion*. In diffusion a gas passes (diffuses) from a region of higher concentration to one of lower concentration.

THE HUMAN RESPIRATORY SYSTEM

The respiratory system includes the lungs and the tubes by which air reaches the lungs. All of the structures of the respiratory system are shown in Figure 7.2.

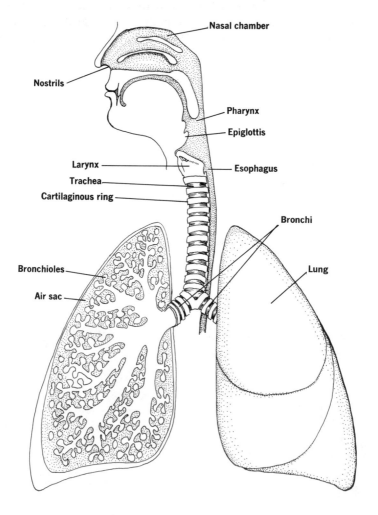

Figure 7.2 The respiratory system.

Air enters the body through the external nares, or *nostrils*, which open into the chamber of the nose (nasal chamber). This is a large cavity above the mouth which contains the sense organs of smell, and is lined with mucus-secreting tissue. Here the air is cleaned of dust particles and warmed before it passes further into the body.

From the nasal chamber the air passes into the throat, or *pharynx*, at the back of the mouth. The pharynx serves as a common passage for both food and air. Food passes from the pharynx to the stomach by way of the esophagus; and air passes from the pharynx to the lungs by way of the larynx, trachea, and bronchi.

The *larynx*, or voice box, contains the vocal cords. Sound is produced by passing air out of the lungs (during expiration) and across folds of tissue. Muscles adjust the tension of this tissue ("cords") to produce the different types of sound and varying pitches possible by the human voice. Only air must enter the larynx. To prevent foods from entering, a flap of tissue, the *epiglottis*, folds over the opening of the larynx whenever food is swallowed. This is done involuntarily.

The other end of the larynx becomes the *trachea*, or windpipe (Figure 7.2). It is distinguished from the esophagus by the "U"-shaped cartilage bands embedded in its wall which hold it open and keep it from collapsing when food is being swallowed down through the esophagus. Also, during inspiration the trachea would collapse without these cartilage bands.

When the trachea reaches the level of the first rib it branches into two cartilaginous tubes (*bronchi*); one goes to the left and one to the right lung. They continue as single tubes until entering the lungs and then each bronchus branches into bronchioles, which in turn continue to branch repeatedly into smaller and smaller tubes that ultimately end in blind sacs—the *air sacs*. The walls of the air sacs are cupped into small cavities termed *alveoli*. These alveoli are covered by a diffuse network of blood capillaries (Figure 7.3). The walls of

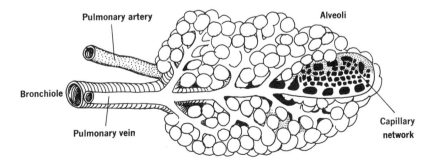

Figure 7.3 The alveoli and capillary network.

the alveoli and the corresponding network of capillaries is where the exchange of oxygen and carbon dioxide between the blood and the external atmosphere takes place (external respiration).

BREATHING

There is a clear distinction between *external respiration* (described above) and *breathing*—which is simply the mechanical process of taking air into the lungs (*inspiration*) and letting it out again (*expiration*).

The lung capillaries are continually removing oxygen from inspired air and adding carbon dioxide to the air of the alveoli. Consequently, this air must be constantly moved in and out of the lungs. In humans this breathing cycle of inspiration followed by expiration is repeated about twelve to eighteen times a minute.

These processes of inspiration and expiration are accomplished by increasing and decreasing the volume of the chest cavity which contains the lungs. During inspiration, the rib muscles (intercostal muscles) and diaphragm contract. When the muscles between the ribs contract they draw the front ends of the ribs upward and outward, increasing the back-to-front area of the chest.

The diaphragm is a sheet of muscle that makes up the floor of the chest cavity. When it contracts it lowers the floor of the chest cavity and enlarges the chest top to bottom. These increases in chest size (front to back and top to bottom) result in a lowering of the air pressure in the lungs. Then, the atmospheric air from outside of the body rushes in (to fill the space) through the nose, pharynx, larynx, trachea, bronchi and its many branches to the air sacs and alveoli. Then, very quickly, the exchange of gases (external respiration) takes place.

At this point the intercostal and diaphragm muscles are fully contracted. To push the air back out of the lungs these muscles relax; this permits the ribs to go back where they were. The diaphragm relaxes, and permits the abdominal organs to push it back up into its previous "dome" shape. This decreases the chest volume and allows the very elastic alveoli and the weight of the chest to push the air out of the lungs.

The air passages (trachea, pharynx, and bronchi) play no active role in the process of breathing. They simply function as air passages.

CAPACITY OF THE LUNGS

A human at rest passes, in and out, only about 500 ml. (milliliter) of air (known as *tidal air*) with each breath. Beyond "at rest" expiration, another 1400 ml. or so of air (*supplemental air*) can be forcibly expelled by fully contracting the abdominal muscles. But, even after the most forcible expiration, there is still about 1200 to 1400 ml. of air (*residual air*) that remains in the lungs which cannot be expelled.

After the normal inspiration of 500 ml., it is possible to inspire deeply, and forcibly take in between 1400 and 3500 ml. more of air (known as *complemental air*). If one forcibly breathes in as deeply as possible and then breathes out as completely as possible he can expel between 3000 and 7000 ml. of air. As shown by Figure 7.4, this amount, which is the combined total of the tidal, complemental, and supplemental air, constitutes the *vital capacity* of the lungs.

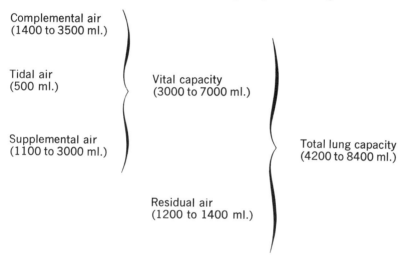

Figure 7.4 Capacity of the lungs. After lungs are filled with air, they are never completely emptied; nor are they ever completely filled.

As mentioned above, even after forcible expiration there is about 1200 to 1400 ml. of residual air in the lungs that cannot be expelled. This air can be expelled only by collapsing the lungs. There is still a small amount of air that remains in the lungs after the lungs collapse; this is the *minimal air*. Taking into account all of the air that is normally found in the lungs, the total capacity of the lungs usually averages between 4200 and 8400 ml. of air.

TRANSPORT OF OXYGEN IN BLOOD

The human body needs about 300 liters of oxygen every twenty-four hours, or one quarter of a liter per minute. But, the human blood circulates about 5 liters of oxygen per minute and supplies more oxygen than needed. When someone starts to do heavy work, run, ski, or place other demands upon the amount of oxygen needed, the blood is able to respond until the demands reach ten or fifteen times the basic requirement. At this point the individual becomes "fatigued."

Approximately 2 percent of the oxygen in the blood is dissolved in the plasma; the rest is carried within the red blood cells, chemically attached to the *hemoglobin*. After the oxygen diffuses into the capillaries of the lungs, it then diffuses into the red blood cells from the plasma and unites with hemoglobin. One molecule of oxygen unites with one molecule of hemoglobin to form a new molecule called *oxyhemoglobin*.

When the oxyhemoglobin-rich blood reaches the capillaries surrounding the cells of the body, the process reverses, reforming the old "reduced" (oxygen-lacking) hemoglobin, which is returned to the lungs for more oxygen.

TRANSPORT OF CARBON DIOXIDE

The transport of carbon dioxide causes a special problem for the body. This is because carbon dioxide dissolves in water and forms carbonic acid. The cells of the body produce, at rest, about 200 ml. of carbon dioxide per minute. If this amount were simply dissolved in the plasma, the amount of carbonic acid produced could cause death. Actually, the carbon dioxide is carried in three different manners. First, an extremely small amount of carbon dioxide is present in the blood as carbonic acid. Second, some carbon dioxide is carried in a loose chemical union with hemoglobin (called carbamino-hemoglobin). Lastly, by far the largest amount of carbon dioxide is converted into bicarbonate compounds through neutralization of the major portion of carbonic acid by sodium or potassium ions, which are released when oxyhemoglobin is changed to hemoglobin.

SUMMARY
   I. Biological oxidation is the process which produces all the energy used within the human body.
      A. Energy producing foods (carbohydrates, proteins, and fats) produce energy by releasing hydrogen to oxygen and forming water.
         1. The oxygen must be brought into the body through the lungs.
         2. The water is used by the body.
      B. Carbon dioxide is also removed from the food during oxidation and must be removed from the body as a waste product.
  II. Respiration
      A. A process with three different meanings.
         1. Cellular respiration is the process by which cells convert food into energy.

2. Internal respiration is the exchange of oxygen and carbon dioxide between the cell and the circulatory system.
3. External respiration is the exchange of oxygen and carbon dioxide through the capillaries of the lungs and air of the external environment.
B. During internal and external respiration the gases diffuse from a region of higher concentration to a region of lower concentration.

III. The Human Respiratory System
A. Includes the lungs and the tubes by which air reaches the lungs.
1. Air enters the body through the external nares, or nostrils, and is passed into the nasal chamber.
2. From the nasal chamber the air passes into the throat, or pharynx.
3. Air passes from the pharynx to the lungs by way of the larynx, trachea, and bronchi.
a. The larynx, or voice box, contains the vocal cords.
b. The trachea is the "windpipe."
c. At the level of the first rib the trachea branches into two tubes termed bronchi; one going to the left and one to the right lung.
(1) Each bronchi branches into bronchioles, which branch repeatedly into smaller and smaller tubes that ultimately end in blind air sacs.
(2) The walls of the air sacs are cupped into small cavities termed alveoli.
(3) The walls of the alveoli and the network of capillaries surrounding them is where external respiration takes place.
B. Breathing is simply the mechanical process of taking air into the lungs (inspiration) and letting it out again (expiration).

IV. Capacity of the Lungs
A. The total capacity of the lungs averages 4200 to 8400 ml.
1. Tidal air is the 500 ml. of air moving in and out of the lungs during normal, "at rest," breathing.
2. Supplemental air is the amount of air that can be forcibly expelled from the lungs beyond the tidal air (about 1400 ml.).
3. Complemental air is the amount of air that can be forcibly breathed into the lungs over and above tidal air, (1400 to 3500 ml.).
4. Vital Capacity
a. The amount of air which a person can breath out by

forcing expiration, after the deepest possible inspiration (3000 to 7000 ml.).
      b. The total of tidal, supplemental, and complemental air.
    5. Residual air is the air in the lungs that can be expired only by collapsing the lungs.
    6. Minimal air is the small amount of air that remains even after the lungs have been collapsed.
V. Transport of Oxygen in Blood
  A. Human blood circulates about five liters of oxygen per minute.
    1. Approximately 2 percent is dissolved in the plasma.
    2. The rest is carried within the red blood cell, chemically attached to the hemoglobin.
  B. At rest, the human body needs about 250 ml. of oxygen per minute.
VI. Transport of Carbon Dioxide
  A. The cells of the body produce, at rest, about 200 ml. of carbon dioxide per minute.
  B. The carbon dioxide is carried to the lungs in three different manners.
    1. An extremely small amount is present in the blood as carbonic acid.
    2. Some is carried in a loose chemical union with hemoglobin (called carbamino-hemoglobin).
    3. The largest amount is converted into bicarbonate compounds and carried to the lungs in this manner.

QUESTIONS FOR REVIEW

1. List the major parts of the human respiratory system and explain how each is adapted for its particular functions.
2. Differentiate between "breathing," "internal respiration," and "external respiration."
3. What is meant by the term "vital capacity?"
4. Explain the capacities of the lungs that go to make up the vital capacity of a person.
5. Describe briefly the sequence of events that take place in the transport of oxygen and carbon dioxide in the blood.
6. Why would someone experience a difficulty in breathing at high altitudes?

Chapter 8
THE DIGESTIVE SYSTEM

Every cell of the body must have a regular supply of chemical nutrients. The digestive system must break down the foods we eat into water-soluble, chemical, molecules small enough to pass from the digestive tract into the blood or lymph and on into the cells of the body. This activity is know as *digestion*. The passage of digested food substances from the digestive tract into the blood and lymph is known as *absorption*.

Digestion involves the action of many *enzymes*. Specific enzymes function in specific areas of the digestive tract; salivary digestion occurs mainly in the mouth; gastric digestion in the stomach; and intestinal digestion in the small intestine. No digestion takes place in the large intestine. This complex process is shown in Figure 8.1.

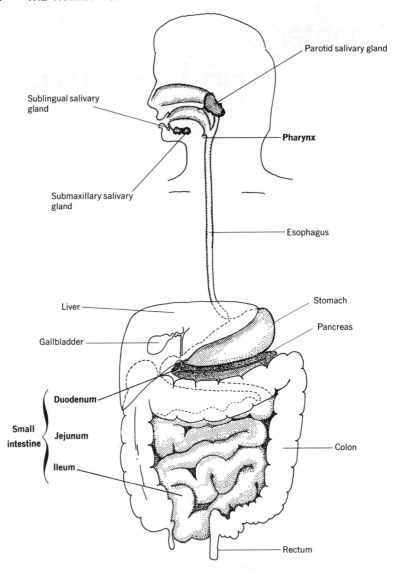

Figure 8.1 The digestive system.

THE DIGESTIVE SYSTEM

The general design of the human digestive tract is a long muscular tube (about twenty-eight feet in length) which begins at the mouth and ends at the anus. The digestive tract consists of the mouth, pharynx, esophagus, stomach, small intestine, and large intestine.

The Oral Cavity

The oral cavity (mouth) begins at the lips and extends into the pharynx. The cavity of the mouth is bounded on the sides by the cheeks; its roof is the palate; and the greater part of the floor is formed by the tongue. The mouth contains the teeth, which reduce food to smaller pieces and thoroughly mix it with the saliva by the action of *chewing* (mastication). Food may then be swallowed as a moistened, soft, round *bolus* (lump) of material.

The Teeth

Teeth are essential for preparing solid food for digestion. Chewing is the most important physical change that food undergoes during digestion. There are four specific benefits obtained from thoroughly chewing food.

1. The food is broken up into smaller particles.
2. Proper mixing of saliva and food enables the digestive juices to work more quickly and thoroughly.
3. The taste of food is enhanced, increasing the pleasure derived from eating. Taste sensations also help to stimulate the secretion of gastric juice (Table 8.1).
4. Blood flow is increased to all of the structures of the mouth. In children such increased blood flow (in conjunction with the factors of heredity, nutrition, and proper tooth eruption) help in the development of the jawbones.

Humans have two sets of teeth (dentitions)—a temporary set and a permanent set (Figure 8.2). A child has twenty-four temporary teeth (also known as deciduous, milk, or baby teeth). These teeth begin to erupt when he is about six or eight months old and continue erupting until he is about twenty-four months old.

The permanent set of teeth gradually replaces the temporary teeth. This process begins during the fifth or sixth year and continues until the seventeenth or even twenty-first year (third molars, or "wisdom teeth"). This permanent set consists of thirty-two teeth. The teeth on each half of each jaw include two incisors (for biting food); one cuspid (for biting and tearing food); two bicuspids or premolars (for grinding food); and three molars (for crushing food).

STRUCTURE OF A TOOTH

The inner portion of a tooth (Figure 8.3) is composed of *dentin*, a soft bonelike material that forms the frame and substance of the tooth. The part of the dentin projecting beyond the jawbone is known as the *crown*, which is covered with a very resistant material, the *enamel*. The section of the dentin embedded into the bone is known as the *root*; it is surrounded by a layer of *cementum*. The ce-

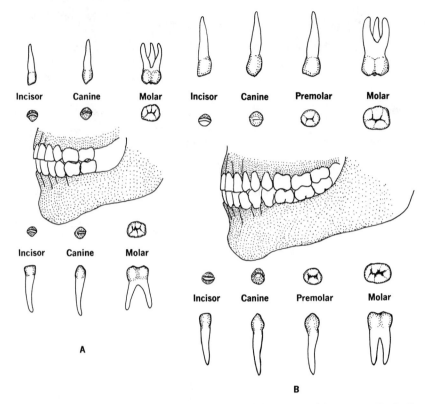

Figure 8.2 The two sets of human teeth. (A) the deciduous or "baby" teeth. (B) the permanent set of teeth.

ment-enamel junction is termed the *neck* of the tooth. Inside the dentin is the *pulp cavity*, consisting of a form of connective tissue; blood vessels; and nerves, which enter the tooth through the opening at the apex (tip) of the root. The root, surrounded by the *periodontal membrane*, is situated in an *alveolar socket* in the bone of the jaw. This periodontal membrane, made up of thousands of tiny fibers, attaches the root to the alveolar socket. It helps to support the tooth in its socket; acts as a cushion by taking up the shock of chewing; and allows for some movement of the teeth. This movement is best illustrated in the field of orthodontia (straightening of teeth), where constant, gentle pressure is applied to move the teeth in the direction desired.

Salivary Digestion

The teeth are instrumental in reducing the food to particles small enough to be swallowed. But, to assist in moving food down the esophagus, as well as to begin the digestive process, a material termed saliva is secreted into the mouth. Surrounding the mouth are three pairs of salivary glands (Figure 8.1). Named according to their loca-

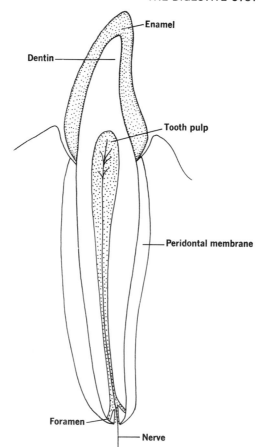

Figure 8.3 Longitudinal section of an incisor tooth in the alveolar socket showing the structure of the tooth and the membranes.

tions (the parotid, the submaxillary, and the sublingual), these glands secrete saliva which contains *mucin* and an enzyme known as *salivary amylase* (ptyalin).

The enzyme functions to change cooked starch into sugar. But food remains in the mouth too short a time for the digestion of starch to proceed very far. This is completed later in the small intestine.

The mucin contained in saliva lubricates the food and assists in swallowing. Saliva also serves to put certain food materials into solution, helping the sense of taste. One can taste only materials which are in solution. Saliva also aids in keeping the mouth and teeth clean.

Continuing on from the mouth is the *pharynx* (throat), which also opens into the nasal cavity. The nose and mouth are separated from one another by the palate, which is hard in the front roof of the mouth and soft in the back. The soft palate projects backwards and reaches almost to the back wall of the pharynx (Figure 8.1).

When a person swallows, the soft palate elevates, preventing the possibility of foods being pushed up into the nasal cavity.

At its lower end the pharynx leads into both the *trachea* (windpipe) and the *esophagus* (food tube). Thus both air and food pass through the pharynx.

The esophagus is a collapsed, muscular tube about ten inches long (Figure 8.1). It is located in the center of the body with its upper two inches in the neck and lower eight inches in the chest cavity (thorax). The esophagus transports food from the pharynx to the stomach. Food is moved through the esophagus by muscular, wavelike contractions called *peristalsis*.

Gastric Digestion

The stomach (Figures 8.1 and 8.4) is an expanded portion of the digestive tract, having an average capacity of about one quart. It is usually in the shape of the letter "J" and lies generally to the left of the midline of the body. The shape of the stomach and the position of its lower part changes from time to time according to the degree to which it is filled with food. The rounded upper part of the stomach is called the fundus, and the middle or main part is the body. The opening through which food passes from the stomach into the intestine is called the *pylorus*. The pylorus is a valve consisting of a circular muscle, which controls movements of fluids or semifluids into the small intestine.

Waves of contractions begin as shallow ripples near the upper end of the stomach and become deeper and stronger as they move toward the pylorus. These muscular waves tend to mix and churn the food. Within two to five hours after a person has eaten, his stomach has reduced much of the food to a partially digested state. This partially liquified material, consisting of food and digestive enzymes (Table 8.1), is called *chyme*. As the stomach completes the mixing, its wavelike action squirts food out through the pylorus into the *duodenum* (Figure 8.4).

Glands line the walls of the stomach and secrete *gastric juice* (Table 8.1). Normal gastric digestive juice is a thin, light-yellow fluid which is a weak acid in infants and children but an increasingly concentrated acid in adults. The acid condition is due to the presence of *hydrochloric acid*. In addition, gastric juice contains various salts and enzymes. The major function of the hydrochloric acid is to provide a medium in which the enzymes can act most completely.

Pepsin (gastric protease) is an enzyme in gastric juice that acts to begin the breakdown of proteins. The fact that the stomach wall, itself a protein, is not digested by pepsin or irritated by the high concentration of hydrochloric acid is attributed to the protective action of the mucus secreted in the stomach.

THE DIGESTIVE SYSTEM

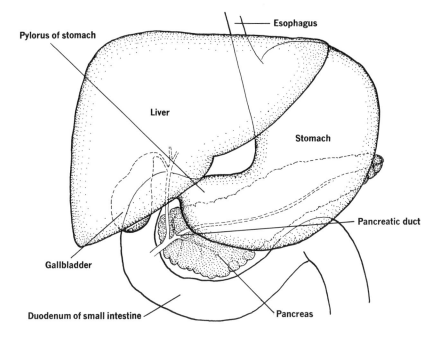

Figure 8.4 Diagram of the relationships of stomach, liver, pancreas, and duodenum.

Table 8.1 The Principal Digestive Enzymes in Humans

ENZYME	GLAND	SUBSTANCE CHANGED	PRODUCTS FORMED
Salivary amylase (ptyalin)	Salivary glands	Boiled starch	Maltose
Gastric protease (pepsin)	Stomach	Protein	Proteoses and peptones
Rennin	Stomach	Casein (milk protein)	Insoluble paracasein (curd)
Gastric lipase	Stomach	Fat	Fatty acids and glycerol
Pancreatic protease (trypsin)	Pancreas	Protein	Proteoses, peptones, and polypeptides
Pancreatic amylase (amylopsin)	Pancreas	Starch	Maltose
Pancreatic lipase (steapsin)	Pancreas	Fat	Fatty acids and glycerol

Table 8.1 (con't.)

ENZYME	GLAND	SUBSTANCE CHANGED	PRODUCTS FORMED
Enterokinase	Intestine	Trypsinogen	Trypsin
Erepsin	Intestine	Peptones and peptides	Amino acids
Maltase	Intestine	Maltose sugar	Glucose sugar
Lactase	Intestine	Lactose sugar	Galactose and glucose sugars
Sucrase	Intestine	Sucrose sugar	Fructose and glucose sugars
Intestinal lipase	Intestine	Fat	Fatty acids and glycerol

Intestinal Digestion

The pylorus of the stomach leads into the small intestine, a tube about twenty-three feet long which is divided into three portions —the *duodenum, jejunum,* and *ileum* (Figure 8.1)

The duodenum (about ten inches long) receives secretions from the pancreas and liver as well as from the intestinal glands which line its own walls. The remaining twenty-two feet of small intestine is made up of the jejunum (the upper two fifths) and the ileum (the lower three fifths). There is no line of distinction between these two, and the division is arbitrary. The main function of these two sections is the absorption of the digested food.

The size of the lumen, the channel within the small intestinal tube, gradually decreases from a diameter of nearly two inches at the upper end of the duodenum to one inch at the lower end of the ileum. The surface area of the membranous lining of the lumen is greatly increased by deep folds (Figure 8.5). These folds are lined with *villi*—microscopic fingerlike projections extending into the lumen. The enormous number of villi give a nubby appearance to the interior surface, which is velvety to the touch. At the base of the villi are the openings of glands which secrete the intestinal digestive enzymes.

PANCREAS

The pancreas is the single most important source of digestive enzymes in the digestive system. It is the second largest gland connected to the digestive tract (the liver is the largest). In the adult the pancreas is approximately six inches long and about one inch wide. It secretes digestive juices into the duodenum by way of the pancreatic duct (Figure 8.4).

The pancreatic juice is a colorless fluid that contains at least

THE DIGESTIVE SYSTEM 103

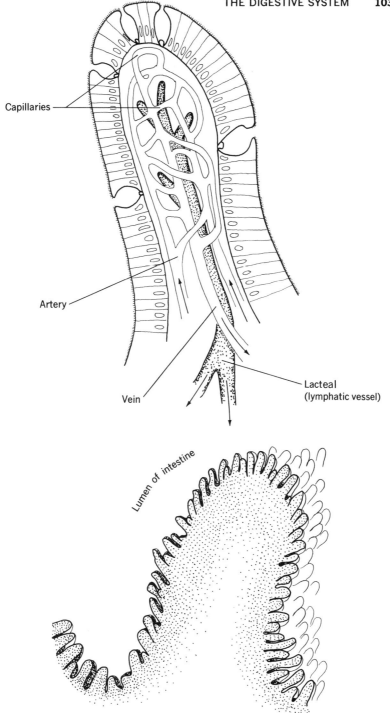

Figure 8.5 Villi of the intestine. Top, the small villi of the intestine. Bottom, a fold in the intestinal wall.

one of every type of enzyme needed to digest food. The major enzymes and their actions are shown in Table 8.1.

The pancreas is also an endocrine gland (see Chapter 11). Certain cells arranged within the pancreas in definite groups, called the *islets of Langerhans,* manufacture the endocrine hormone *insulin.* As explained in Chapter 11 this hormone plays an important role in the control of carbohydrate metabolism of the body.

LIVER

The liver is the largest organ of the body (Figure 8.4). It is situated mainly on the right side of the abdomen, lying under and protected by the lower ribs. The liver is the great chemical laboratory of the body and is important in the metabolism of carbohydrates, proteins, and fats.

The digestive function of the liver is to secrete *bile,* which passes into the intestine after storage in the gallbladder. At mealtime bile flows through a duct from the gallbladder into the duodenum (Figure 8.4). At other times a muscular valve prevents its passage.

Bile is not an enzyme but contains many salts, fatlike substances, and pigments. It is a yellow-brown or greenish fluid. The pigments are responsible for its color and produce the color characteristic of *feces,* the body's solid wastes. The bile salts are the active constituents of bile. They act as wetting agents, lowering the surface tension of the fatty film surrounding fat particles in food. This process is termed *emulsification.* It enables the fats to undergo division into smaller globules and form a fine emulsion, which facilitates closer contact between the digestive enzymes and the fat particles. The digestive action of the enzymes is increased many times by this close contact.

INTESTINAL JUICE

The intestinal juice contains a large number of enzymes that complete the digestive process. The most important of these enzymes and their individual actions are outlined in Table 8.1.

Intestinal Absorption

The absorption of certain drugs occurs through the membranes of the mouth. Small amounts of water, salts, glucose, and alcohol can be transferred through the membranes of the stomach into the blood stream. Nearly all other digested food is absorbed through the membranes of the small intestine.

The villi (Figure 8.5) are the structures responsible for the absorption of digested food from the lumen of the small intestine. Each of the 4 to 5 million villi in an adult contains a central lymph vessel called a *lacteal,* surrounded by a network of blood capillaries (Figure 8.5). After food has been digested, it passes through the walls of the intestine into the capillaries and lacteals of the villi.

Simple sugars (monosaccharides) and amino acids pass into the capillaries. Fatty acids and glycerol (end products in the digestion of fats) may either pass directly into the capillaries, or, once they have passed through the wall of the intestine, recombine and pass into the lacteals.

Other significant materials which are absorbed in the small intestine are vitamins, water, and salts. The fat-soluble vitamins A and D (Table 8.2) are absorbed along with dietary fats. The water-soluble vitamins, B, C, and the others, are readily absorbed through the capillaries. Water and salts are similarly absorbed.

Large Intestine (Colon)

The small intestine opens into the first part of the large intestine, the *cecum* (Figure 8.6). The entrance into the large intestine is through a muscular valve, the *ileocecal valve,* which prevents the backflow of the cecal contents into the ileum. Also opening into the cecum is a blind sac, the *vermiform appendix,* which ranges in length from two to eight inches. The functions of the appendix are not known.

As shown by Figure 8.1, the large intestine extends up the right side of the abdomen as the *ascending colon,* across the abdomen as the *transverse colon,* and down the left side of the abdomen as the *descending colon.* After making an S-shaped curve called the *sigmoid colon,* the large intestine terminates as the *rectum.* The rectum is about six inches long and opens to the exterior of the body through the *anus.* The retention of the rectal contents is controlled by the sphincter muscles (ringlike muscles) which are constantly active and contracted except during evacuation of the waste product, the feces. The act of excreting feces from the rectal reservoir is known as *defecation.*

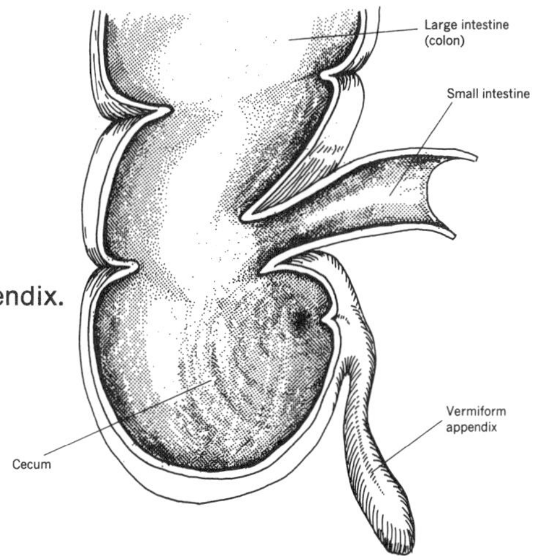

Figure 8.6 Cecum and appendix.

Table 8.2 Vitamins

VITAMIN	RICH SOURCES	PROPERTIES	FUNCTION	DEFICIENCY SYMPTOMS
FAT-SOLUBLE VITAMINS				
Vitamin A	Cheese, green and yellow vegetables, butter, eggs, milk, fish liver oils; carotene in vegetables converted to vitamin A by liver	Lost through oxidation during long cooking in open kettle; overdose possible	Necessary for growth, tooth structure, night vision, healthy skin	Slow growth, poor teeth and gums, night blindness, dry skin and eyes (lack of tears)
Vitamin D	Beef, butter, eggs, milk, fish liver oils; produced in the skin upon exposure to ultraviolet rays in sunlight; no plant source	One of the most stable vitamins; large doses may cause calcium deposits, poor bone growth in children, congenital defects	Necessary for metabolism of calcium and phosphorus; essential for normal bone and tooth development	Rickets; poor tooth and bone structure; soft bones
Vitamin E	Widely distributed in foods; abundant in vegetable oils and wheat germ	Lost through oxidation during long cooking in open kettle; overdose not known	Not definitely known for humans	Not definitely known for humans
Vitamin K	Eggs, liver, cabbage, spinach, tomatoes; produced by bacteria of intestine	Destroyed by light and alkali; absorption from intestine into blood depends upon normal fat absorption	Necessary for blood clotting	Slow blood clotting; anemia

WATER-SOLUBLE VITAMINS

Vitamin	Sources	Characteristics	Function	Deficiency
Vitamin B_1 (thiamine)	Meat, whole grains, liver, yeast, nuts, eggs, bran, soybeans, potatoes	Not destroyed by cooking, but being water-soluble, may dissolve in cooking water; not stored in body; daily supply needed	Necessary for carbohydrate metabolism, normal nerve function; promotes growth	Beriberi; slow growth, poor nerve function, nervousness, fatigue, heart disease
Vitamin B_2 (riboflavin)	Milk, cheese, liver, beef, eggs, fish	Not destroyed by cooking acid foods; unstable to light and alkali	Essential for metabolism in all cells	Fatigue, sore skin and lips, bloodshot eyes, anemia
Niacin (nicotinic acid)	Bran, eggs, yeast, liver, kidney, fish, whole wheat, potatoes, tomatoes; can be synthesized from amino acid tryptophan	Not destroyed by cooking, but may dissolve extensively in cooking water	Necessary for growth, metabolism, normal skin	Pellagra; sore mouth, skin rash, indigestion, diarrhea, headache, mental disturbances
Vitamin B_6 (pyridoxine)	Meat, liver, yeast, whole grains, fish, vegetables	Stable except to light	Functions in amino-acid metabolism	Dermatitis; deficiency rare
Vitamin B_{12} (cyanocobalamin)	Meat, liver, eggs, milk, yeast	Unstable to acid, alkali, light	Necessary for production of red blood cells and growth	Pernicious anemia
Vitamin C (ascorbic acid)	Citrus fruits, tomatoes, potatoes, cabbage, green peppers, broccoli	Least stable of the vitamins; destroyed by heat, alkali, air; dissolves in cooking water	Essential for cellular metabolism; necessary for teeth, gums, bones, blood vessels	Scurvy; poor teeth, weak bones, sore and bleeding gums, easy bruising, poor wound healing

NOTE: Several other water-soluble vitamins are believed to be essential to human nutrition, but are not as well understood as the above vitamins and their deficiency is less common.

Though no digestive enzymes are secreted by the large intestine, bacteria break any unabsorbed amino acids into simpler compounds and gases, many of which have strong odors and may be toxic. The odor of feces is due mainly to this process. The color of the fecal material is caused by the action of bacteria on the bile pigments. As well as these effects, the bacterial action of the intestine performs an important function in nutrition. It synthesizes vitamin K and certain vitamins of the vitamin B complex group, which are then absorbed into the blood stream through the walls of the large intestine.

One of the principal functions of the large intestine is the absorption of water. Material entering into the cecum is fluid, but by the time the undigested waste has reached the lower portion of the descending and sigmoid colon its consistency is that of paste, due to the loss of water. Little if any digestible food remains in the feces. In other words, almost all of the protein, fat, and carbohydrate that is eaten is absorbed; the food residue of the feces consists almost entirely of indigestible substances. Vegetable material, since its framework is composed of indigestible cellulose, contributes more bulk to the feces than do other foods. This indigestible material, or "roughage," serves a useful purpose in that it acts as a mechanical stimulus, increasing the mobility of the ingested material as well as the secretions from the intestinal wall. Fecal material consists of bacteria (9 percent or more); solids (mainly nitrogenous wastes); and minerals excreted into the large intestine from the blood, together with loose cells and white blood cells shed from the intestinal lining.

SUMMARY
I. Every cell must have a regular supply of chemical nutrients.
   A. Digestion is the breakdown of foods into water-soluble, absorbable chemicals.
   B. Absorption is the passage of digested food substances from the digestive tract into the blood and lymph.
II. The Digestive system
   A. The digestive system is a long muscular tube which begins at the mouth (oral cavity) and ends at the anus.
   B. The Oral cavity (mouth)
      1. Begins at the lips and extends into the pharynx.
      2. Contains the teeth.
      3. Salivary digestion takes place within the oral cavity.
   C. The Teeth
      1. Reduce food to smaller pieces and thoroughly mix it with the saliva by the action of chewing (mastication).
      2. Humans have two sets of teeth—a temporary set and a permanent set.

3. Structure of a tooth shown in Figure 8.3.
D. Salivary digestion
   1. Teeth are instrumental in reducing the food to particles small enough to be swallowed.
   2. Salivary glands in the mouth secrete saliva which contains mucin and salivary amylase.
      a. Salivary amylase changes cooked starch into sugar.
      b. Mucin lubricates the food and assists swallowing.
   3. Saliva also serves to put certain food materials into solution, helping the sense of taste.
   4. Saliva also aids in keeping the mouth and teeth clean.
   5. The mouth enters into the pharynx which leads into both the trachea (windpipe) and the esophagus (food tube.)
   6. The esophagus transports food from the pharynx to the stomach by muscular, wavelike contractions called peristalsis.
E. Gastric digestion
   1. The stomach is an expanded portion of the digestive tract having an average capacity of about one quart.
   2. The stomach is shaped like the letter "J" and lies to the left of the midline of the body.
   3. Food is partially liquified, mixed with digestive enzymes, and moves out through the pylorus into the duodenum as chyme.
   4. Gastric juice is a thin, light-yellow fluid which is weakly acid in infants and children but increases its acidic concentration in adults. It contains:
      a. Hydrochloric acid—which accounts for the acidic condition.
      b. Pepsin (gastric protease), an enzyme that acts to begin the breakdown of proteins.
F. Intestinal digestion
   1. The pylorus of the stomach leads into the small intestine.
   2. Small intestine is divided into three portions—the duodenum, jejunum, and ileum.
      a. The duodenum is the shortest portion; into it flow the secretions from the pancreas and liver.
      b. The remaining portion of small intestine is made up of the jejunum and the ileum—these mainly function as areas for the absorption of the digested food.
   3. Pancreas
      a. Single most important source of digestive enzymes in the digestive system.
   4. Liver
      a. Digestive function of the liver is to secrete bile.

b. Bile is not an enzyme.
   c. Bile acts as a wetting agent, lowers the surface tension of the fatty film surrounding fat particles in food enabling the fats to be more easily digested.
  5. Intestinal juice contains a large number of enzymes that complete the digestive process.
G. Intestinal absorption
  1. Certain drugs are absorbed through the membranes of the mouth.
  2. Small amounts of water, salts, glucose, and alcohol can be absorbed through the membranes of the stomach.
  3. Nearly all other digested food is absorbed through the membranes of the small intestine.
H. Large intestine (colon)
  1. The small intestine opens into the large intestine at the cecum.
  2. The large intestine
   a. Extends up the right side of the abdomen as the ascending colon.
   b. Across the abdomen as the transverse colon.
   c. Down the left side of the abdomen as the descending colon.
   d. Makes an S-shaped curve called the sigmoid colon.
   e. Terminates as the rectum.
  3. No digestive enzymes are secreted by the large intestine.
  4. Bacteria break any unabsorbed amino acids into simpler compounds and gases.
  5. Bacteria also synthesize vitamin K and certain vitamins of the vitamin B complex group, which are then absorbed into the blood stream.
  6. Principal function of the large intestine is the absorption of water.
  7. The unabsorbed, indigestible food residue builds up in the rectum as feces. The act of evacuation of this waste material is known as defecation.

QUESTIONS FOR REVIEW
1. List the organs of the human digestive system in order and describe the functions performed by each.
2. Draw a tooth and label its main parts.
3. Describe the usefulness of chewing in digestion.
4. What keeps the food moving through the digestive tract?
5. Explain the function of the stomach. What are the functions of gastric digestion?

6. Discuss the role of enzymes in digestion.
7. What is the role of bile in digestion? Where is it produced? How, and under what conditions, does it reach the food undergoing digestion?
8. Distinguish between digestion and absorption. Absorption takes place in which areas of the digestive system?
9. Explain the term "roughage." What function does this material serve in digestion?
10. What are "feces?" What is the composition of fecal material?

Chapter 9
THE EXCRETORY SYSTEM

Food and oxygen are essential for growth, repair, synthesis of hormones and other body chemicals, and energy for all cells of the body. Food is supplied to the blood stream by the digestive organs and oxygen is supplied by the lungs. But, as a result of all of these necessary chemical activities within the body, waste products are formed which can affect the state of affairs in and around the cells. Unless the level of these wastes is kept within a normal range, cell functioning slows down and the death of cells (and even the individual) can occur. If working properly, the organs of excretion can effectively eliminate these wastes.

MATERIALS TO BE ELIMINATED

Wastes to be eliminated include water, carbon dioxide, products from the breakdown of proteins (such as urea and creatinine), and heat. Also present are microorganisms (dead and alive) found in large numbers in the feces.

The substances to be eliminated are known as *excreta*, and the process by which they are eliminated from the body as *excretion*, or elimination.

This chapter will describe the urinary system in detail. Excretion from the digestive tract has been discussed in Chapter 8. Removal of wastes by the skin and lungs has been discussed in Chapters 5 and 7 respectively and will not be further included here.

THE URINARY SYSTEM

The urinary system consists of the two *kidneys*, which extract the urine from the blood; two *ureters*, which transport urine to the *urinary bladder*; the urinary bladder; and the *urethra* which carries the urine out of the body (Figure 9.1).

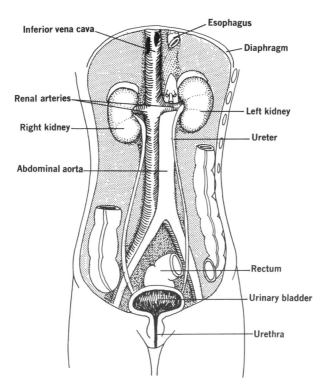

Figure 9.1 Urinary structures, front view showing gross anatomy.

The Kidneys

LOCATION AND EXTERNAL STRUCTURE

As seen in Figure 9.1, the kidneys consist of two bean-shaped organs located in the back of the abdomen. Anatomically this is directly in front of the last thoracic and first three lumbar vertebrae (you may wish to also refer to Figure 1.1 of the skeletal system). Because the liver occupies the upper right hand corner of the abdominal cavity, the right kidney must fit under it and so is a little lower than the left one.

Each kidney is embedded in thick fatty (*adipose*) tissue. This mass of fat holds each kidney in place and protects it from mechanical damage. This is important to the kidney because there are no ligaments to hold it in place. Around the kidney, and within the bed of fat, is a layer of tough fibrous tissue called *renal fascia* which further supports the kidney. If the kidney slips out of place it is described as a *floating kidney*.

Each kidney is shaped like an elongated oval with a notch taken out of one side. The area of the notch is called the *hilum* and this side is always turned toward the center of the body (Figure 9.1). The top of each kidney is surrounded by an adrenal gland. Each kidney weighs, on the average, about one-third of a pound and is about four to five inches long and two to three inches wide.

Passing through the hilum are blood vessels, nerves, and the *ureter* (the duct between the kidney and bladder). The largest of these are: (1) the *renal artery*, (2) the *renal vein*, and (3) the *ureter*. The hilum leads into the *renal sinus* (large cavity within the kidney). It might be interesting to note that the word *renal* is a scientific word which pertains to the kidney. Wherever renal is used, the word kidney could be correctly substituted.

INTERNAL STRUCTURE

If the kidney is cut in two, lengthwise, the blood vessels and nerves entering the renal sinus can easily be seen. A good deal of the space within the sinus is occupied by the *renal pelvis*, a funnel-shaped reservoir which connects to the ureter. Entering the pelvis from the tissue of the kidney are a number of small ducts. Each of these ducts is a *calyx* (Figure 9.2).

The mass of the kidney tissue can be divided into an outer *cortex* and an inner *medulla*. The medulla is divided into cone-shaped structures called *renal pyramids*. The pyramids are wide toward the cortex, and narrow toward the sinus. The cortex not only covers the base of each pyramid, but also dips down between them forming the *renal columns*.

The bulk of the cortex and medulla consists of about one mil-

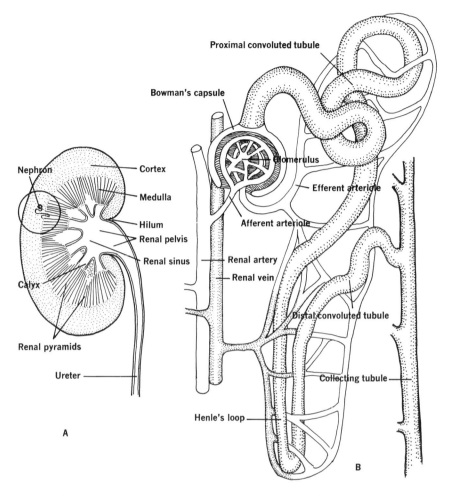

Figure 9.2 Kidney. (A) internal structure of kidney. Insert shows location of nephron showing renal circulation.

lion small tubes, or renal tubules, packed closely together side by side. Each tubule and its blood supply is called a *nephron* (Figure 9.2).

Each nephron begins somewhere in the cortex as an expanded, cuplike structure called a *Bowman's capsule* and then goes through a series of twists. From the Bowman's capsule, the tube leads to the twisted *proximal convoluted tubule*, then to the *Henle's loop* which dips into the medulla, then to the twisted *distal convoluted tubule*, which finally leads into a *collecting tubule*. Each collecting tubule descends into a renal pyramid (in the medulla) which opens into a renal calyx which flows into the renal pelvis.

The function of each nephron is closely associated with the circulation of the blood. The kidney receives a rich blood supply from the renal artery. This artery subdivides into smaller and smaller branches called *afferent arterioles* (Figure 9.2). Each afferent arteriole moves into the cuplike Bowman's capsule. Here the arteriole breaks up into a capillary tuft called a *glomerulus*. Then the glomerulus empties into another arteriole, the *efferent arteriole*. As the efferent arteriole moves away from the capsule, it moves toward the tubule, where it breaks up into a capillary network which twines around the convoluted tubules and Henle's loop. From this network it flows into the renal vein. The essence is that a great deal of blood flows around each nephron.

The Ureters

Each *ureter* is about twelve inches long and one-half inch in diameter. Each one extends from the renal pelvis to the bladder. The opening of the ureter into the bladder is in the form of a little slit which prevents the backflow of urine. When the bladder is empty, the slit opens whenever drops of urine move down the ureter. When the bladder is full, the slit stays closed and will not allow more urine into the bladder. In this case urine backs up the ureter and into the kidney.

The Urinary Bladder

The average human urinary bladder can comfortably hold about one pint of urine. It lies right behind the symphysis pubis (see the discussion on "Legs and Pelvic Girdle" in Chapter 1). In the male the bladder contacts the rectum and in the female it is next to the uterus and vagina. When the bladder is full it can rise some distance in the abdominal cavity.

The Urethra

The *urethra* is a tube which transports the urine from the bladder to the outside of the body. It differs in men and women.

MALE URETHRA

It is a good idea at this point to refer to Figure 10.1 in Chapter 10. The urethra in the male serves to transport urine and to ejaculate semen. The two sperm ducts enter the urethra through the prostate gland. The urethra passes through the penis and terminates in a slitlike opening. The male urethra is about eight inches long.

FEMALE URETHRA

Now refer to Figure 10.2 in Chapter 10. The female urethra is much shorter than that of the male (about one inch long) and exits between the clitoris and the vaginal opening.

THE FORMATION OF URINE

Now we can examine the way in which the kidney produces urine. There are two phases to urine production; the first is *filtration*, and the second is *reabsorption*.

Filtration

As the blood flows through the glomerulus, many blood components escape into the Bowman's capsule. This is called *filtration*, and is due largely to blood pressure. Among these components are blood sugar (glucose), blood plasma, amino acids, and urea. In fact, everything escapes *except* blood corpuscles and huge blood proteins—both of which are too large to make it. The amount of fluid escaping into the nephrons of both kidneys is about *forty-four gallons per day!* Of course, all of this is not excreted as urine. Almost all of the forty-four gallons consists of usable materials which must not be wasted.

Tubular Reabsorption

As the fluid moves down through the convoluted tubules and Henle's loop, a great deal of it is reabsorbed into the blood in the capillaries. The body needs the glucose, so it is returned to the blood. The same is true for much of the water, sodium, and certain other materials. After reabsorption is completed, the product that remains is the *urine*. The urine then passes from the collecting tubule into the renal pelvis and on to the bladder. The amount of urine excreted seldom amounts to more than two and one-half quarts per day.

Volume and Composition of Urine

Actually, the daily volume of urine eliminated from the normal adult will vary from one to two quarts (and may even be less on occasion). This fluctuation depends on such things as water intake, air temperature, the food we have eaten, the sort of physical activity we have done, and our emotional state.

Not all body fluid is lost through the kidneys. Considerable amounts are also lost through the skin. When the body is exposed to higher temperatures, the skin perspires freely; when exposed to cold, sweat production is stopped almost completely. The more body water lost through perspiration, the less through urine and vice versa. The urine volume is lower in the summer and in warm climates, and higher in the opposite situations.

When a person drinks large quantities of liquids, urine volume increases. Certain substances taken in with food and water may also increase urine flow. These substances are called *diuretics* and include such things as coffee, tea, and alcoholic beverages. Coffee contains caffeine, which chemically stimulates urine production. Of course, the water in the coffee means increased volume as well. Alcohol acts in somewhat the same way. Drinks of low alcoholic content will cause little more urine output than drinking similar amounts of water. However, beverages of higher proof exert their diuretic effect principally because of higher alcoholic content and not the volume of liquid.

Urine is usually a yellow (amber) transparent liquid with a characteristic odor. The color of urine will vary with the volume of water consumed, the particular sorts of food in the diet, certain drugs, and the presence of certain diseases. Normally urine should be transparent and not cloudy. Cloudiness may simply reflect a person's diet, but in some cases it may denote abnormal, or diseased conditions.

URINATION

As the bladder fills with urine, the pressure inside the bladder remains about the same. At about one-third of a quart (in the adult) there may be a sensation of fullness and a desire to urinate. This feeling may be suppressed. However, when two-thirds of a quart, or more, collect, a feeling of pain may develop and the urge to urinate becomes irresistible.

When enough urine collects in the bladder, nerve impulses cause the muscles in the bladder wall to contract and the ones guarding the urethra to relax, allowing urination. It is possible, however, to control the flow by straining one's abdominal muscles. Very young children have poor control over muscles regulating the urethra—as is apparent to any mother whose child wets his bed at night. Improved control develops with age and training.

Abnormal Conditions

Dead cells, pus, albumin, blood, and glucose are some things found in urine under certain conditions. Specific terms used to describe the presence of some of these substances would be: for blood, *hematuria*; glucose, *glycosuria*; and albumin, *albuminuria*.

Frequent or continuous presence of glucose in the urine indicates excessively high amounts of sugar in the blood. As occurs in *diabetes mellitus*, the kidneys are not able to reabsorb all of the sugar. Blood in the urine may indicate infection, cancer, or kidney stones. High levels of mineral salts in the urine may lead to the formation of kidney stones in the kidney or ureter. These may be washed out by the urine or removed surgically. Their formation may be caused by inadequate water consumption, too much blood salts, or other causes. Abnormal urine production, either in appearance or volume, should be referred to a physician.

SUMMARY
I. Materials to be eliminated, known as excreta, include water, carbon dioxide, protein break-down products, and heat.
II. The Urinary System
   A. The kidneys
      1. Location and external structure
         a. In the back of the abdomen.
         b. Elongated oval shape with notch out of one side.
      2. Internal structure
         a. Renal pelvis is funnel-shaped reservoir which connects to the ureter.
         b. Outer region of solid tissue is called cortex; inner region, medulla.
         c. Cortex and medulla consist mainly of nephrons, each including the following parts:
            (1) Bowman's capsule.
            (2) Proximal convoluted tubule.
            (3) Henle's loop.
            (4) Distal convoluted tubule.
            (5) Glomerulus.
            (6) Afferent and efferent arterioles.
         d. Collecting tubules carry urine from nephrons to renal pelvis.
   B. The ureters
      1. Tubes about twelve inches long.
      2. Carry urine from kidney to bladder.
   C. Urinary bladder holds about a pint of urine.
   D. Urethra carrys urine from bladder to outside of body.
      1. Male urethra about eight inches long, passing through penis; carries semen as well as urine.
      2. Female urethra only about one inch long.
III. Formation of Urine
   A. Filtration
      1. From the blood in the glomerulus into Bowman's capsule.
      2. Includes all components of blood except cells and proteins.
      3. Amounts to about forty-four gallons per day.
   B. Reabsorption
      1. From the convoluted tubules and Henle's loop back into the blood in the capillaries.
      2. Most of the water and dissolved materials in the filtrate returns to the blood.
      3. Product that remains is the urine.
      4. Volume of urine seldom exceeds two and one-half quarts per day.

C. Volume and composition of urine
   1. Depends on water intake, air temperature, diet, activity level, and emotional state.
   2. Volume is reduced by increased sweating.
   3. Caffeine and alcohol increase urine volume, acting as diuretics.
   4. Color and clarity of urine affected by many factors.
D. Urination
   1. Feeling of fullness when about one-third full.
   2. Feeling of discomfort and irresistible urge to urinate when about two-thirds full.
   3. Nerve impulses cause muscles of bladder wall to contract and muscles of urethra to relax causing urination.
   4. Young children prone to bed wetting due to poor control of urethral muscles.
E. Abnormal conditions
   1. Hematuria—blood in urine.
   2. Glycosuria—glucose (sugar) in urine.
   3. Albuminuria—albumin (protein) in urine.

QUESTIONS FOR REVIEW

1. What organs, other than those of the urinary system, function in excretion?
2. What is the function of the nephron?
3. Contrast the male and female urethra.
4. Name and discuss the phases in urine formation.
5. What factors may influence the volume and composition of urine?
6. Define hematuria, glycosuria, and albuminuria.

Chapter 10
THE REPRODUCTIVE SYSTEMS

Many tissues of the body undergo regeneration, or replacement during the life of a person. Worn-out skin cells are shed, fractured bones are fused, and old blood cells are replaced. Such regeneration, however, does not apply to organs, large body parts, or the whole person. A severed arm is never replaced during a person's life-time; a dead man is not brought back to life.

The only way in which a whole individual can be reproduced is through sexual reproduction. Sex cells grow and develop within reproductive organs in both male and female. The male sex cell fertilizes the female egg cell, and that cell develops through repeated cell divisions into an adult individual.

THE MALE REPRODUCTIVE SYSTEM

Production of Sperm

TESTES

The male gonads are responsible for the production of sperm and hormones. The two gonads, also known as the *testes* or *testicles*, are oval glands about one and one-half inches long. They are suspended from the underside of the body in a bag, the *scrotum* (Figure 10.1). Sperm production can not take place at the normal body temperature. The scrotum permits the testes to suspend from the body and thus to maintain a temperature three to four degrees lower than normal body temperature, an optimum temperature for sperm production. Exposing the testes to high temperatures may induce temporary infertility; prolonged exposure to high temperatures may induce sterility. Cold temperatures inhibit sperm production, but do not destroy sperm cells. Egg production in the female is not affected by normal body temperature.

The testes also serve as *endocrine glands* which produce the male sex hormones. These hormones begin to flow in large amounts when the male is about thirteen years of age. This action marks the

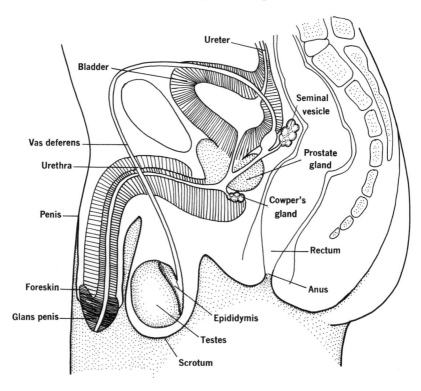

Figure 10.1 Male reproductive system, side view.

onset of *puberty*, or the beginning of the physical changes leading to sexual maturity. This process also leads to the development of broader shoulders, lowered voice, and hair on the face, chest, and pubis. These physical developments are called *secondary sex characteristics*.

SEMINIFEROUS TUBULES

Within each testis are a great number of very small tubes called *seminiferous tubules*. Within the lining of each tubule the sperm cells are formed. Starting at puberty, these sperm cells are produced without interruption throughout the lifetime of the male. At first they are produced in small numbers. Gradually the number increases, until the sexually adult male produces ten to thirty billion sperms each month. The total number of sperm cells produced during a man's life-time is almost impossible to determine.

Sperm cells are not only great in number but also exceedingly small. Each cell is microscopic, about fifty microns long. It consists of a *head, neck, body* and *tail,* and carries within its head, one set of chromosomes (23).

EPIDIDYMIS

As the sperm mature they pass into a coiled tube at the back of each testis, the *epididymis*. In this reservoir they are stored, inactive, in a thick liquid. They may remain there until released from the body, or they may disintegrate and be reabsorbed by the tubules.

SEMINAL DUCTS

The sperm are carried out of the scrotum through two small tubes called the *seminal ducts (vas deferens)*. These ducts extend from each epididymis up into the pelvic cavity and empty into the *urethra*, which is beneath the urinary bladder. The urethra serves as a passageway for sperm and urine (the two, however, will never be discharged at the same time).

Fluid-producing Glands

Sperm moving from the testes through the seminal ducts unite with fluids produced by a series of glands (the two *seminal vesicles* and the *prostate*). One seminal vesicle empties into each seminal duct near the point where they unite to become the urethra. The prostate is located just below the urinary bladder, and surrounds the point where the two seminal ducts merge into the urethra. Farther down the urethra is another pair of small glands, the *Cowper's glands,* which also produce a fluid. The secretion from *Cowper's glands* appears as a drop of clear fluid at the end of the penis just prior to ejaculation. It removes any urine which may be in the urethra and lubricates the penis.

All of these secretions, together with the sperm cells, are called *semen* or *seminal fluid*. Seminal fluid is a thick, whitish, alkaline material. When mixed with it, the sperm gain the ability to move by themselves.

Penis

The semen is ejected through the *urethra*, which extends outside of the body to the end of the *penis*. Besides conducting the urine to the outside of the body, the penis also introduces the semen into the genital passages of the female so that fertilization can take place. In order to accomplish this, the penis, which is usually limp, must become firm and erect, a state called *erection*.

The tissue of the penis is composed of many large blood vessels that are empty of blood when the penis is limp, but engorged with blood during erection. The globular end of the penis is called the *glans penis* (Figure 10.1). It is richly supplied with nerve receptors, making it especially sensitive to external stimulation. An erection may be brought on by physical manipulation of the penis, by sexual thoughts, by pressure from a full urinary bladder or rectum, from wearing clothing which fits too tightly, or from anything that causes congestion of blood in the region of the penis.

The inability to attain an erection is called *impotence*. It is not to be confused with *sterility*, which is caused by either an insufficient number or total lack of sperm cells. A man may be sterile, yet fully potent.

Between the body of the penis and the glans penis there is a groove. From the body of the penis there is normally a free fold of skin, the *foreskin* or *prepuce*, which overhangs the glans penis when the penis is limp. Circumcision, which means "cutting around," is an operation in which an incision is made around the penis to remove all or a part of the foreskin.

EJACULATION

Physical stimulation of the penis not only causes it to become erect, but finally results in a forcible expulsion of semen, called *ejaculation*. The muscles lining the walls of the seminal ducts contract and push the sperm cells into the urethra. At the same moment the fluid-producing glands contract and secrete their fluids. Then, by a series of wavelike contractions, the semen is discharged through the penis. In sexual intercourse, ejaculation occurs at the climax and sperm cells are placed in the female reproductive tract.

At the time of ejaculation, approximately four milliliters (4 ml.) of semen are released. The average number of sperm in man is 120 million sperm per milliliter of semen. Thus more than half a billion sperm are released in a normal ejaculation, although only a single sperm is required to fertilize the egg. However, good fertility is not assured if the sperm count falls below 50 million per milliliter. Also, if as many as 25 percent of the available sperm are abnormal in form, the man may be sterile.

Ejaculation is usually accompanied by a feeling of intense sexual excitement and emotional release called *orgasm*. Shortly after ejaculation the orgasm subsides, the penis becomes limp, and the male feels sexually satisfied.

Male Reproductive Hormones

The principal hormone produced by the testes is *testosterone*. It is formed by cells between the seminiferous tubules called *interstitial cells*. It is responsible for the development of the male secondary sex characteristics as well as the development of the reproductive organs.

The production of testosterone is, in turn, under the direct control of two hormones secreted by the *pituitary gland*, which is situated at the base of the brain. The two hormones, *follicle-stimulating hormone* (FSH) and *interstitial-cell-stimulating hormone* (ICSH), are described as gonadotropins, since they affect the gonads. FSH causes the seminiferous tubules to produce sperm cells. ICSH causes interstitial cells to produce testosterone. When the amount of the testosterone becomes too great, it reacts against the ICSH, which is then reduced. Testosterone can also serve to control the FSH.

The absence or removal of the testes may cause hormonal deficiencies in the male. Since the production of testosterone become increasingly important to the male after the time of puberty, the effects of a testosterone deficiency depend on when it occurs. If it occurs before puberty, a boy fails to develop male secondary sex characteristics. A boy who loses his testes prior to puberty is known as a *eunuch*. If the loss occurs after the time of puberty, he will retain some male secondary sex characteristics and lose others. The removal of the testes is called *castration*.

THE FEMALE REPRODUCTIVE SYSTEM

Ovaries

In the female, the production of sex cells (*eggs* or *ova*) is accomplished by the *ovaries*. The two ovaries in the female correspond to the two testes in the male; they may also be called gonads. In contrast to the location of the testes in the male, the ovaries in the female are situated within the pelvic cavity (Figure 10.2).

Like the male gonads, the ovaries perform a dual function

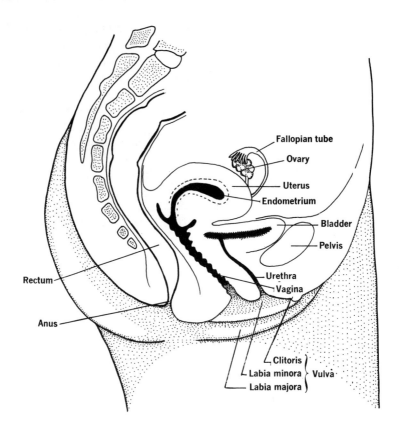

Figure 10.2 Female reproductive system, side view.

and produce both egg cells and hormones. At birth, the ovaries contain about 400,000 immature egg cells. When a girl reaches puberty, these underdeveloped eggs begin to mature, and the ovaries begin to release the hormones *estrogen* and *progesterone* into the blood stream.

One of these hormones, estrogen, is primarily responsible for causing the enlargement of the breasts and the hips and the development of a feminine appearance (the *secondary sex characteristics*).

Each immature egg is enclosed in a minute bubble, or *immature follicle*. Beginning with puberty, these immature follicles ripen or mature at the approximate rate of one each month and develop into a *graafian follicle*. Once each month, usually midway between menstrual discharges, a follicle ruptures and releases a mature ovum.

Fallopian Tubes

Each ovary is adjacent to the fringed (fimbriated) ends of the *fallopian tubes* or *oviducts*, which lead to the uterus. Once the ovum is released from the ovary, it enters one of the fallopian tubes (Figure 10.3).

The inner lining of the fallopian tube is covered with minute, hairlike structures called *cilia*. Once the egg is released from the follicle, it is swept into the fallopian tube and carried downward toward the uterus by the action of the cilia.

The expulsion of the egg from the ovary is called *ovulation*. At this stage the egg, if it could be viewed, would just barely be visible to the unaided eye. Once the egg is released from the ovary, it may unite with (be *fertilized* by) any sperm which may be present. Three days are normally required for the transport of the egg through the fallopian tube into the uterus. Since the released egg is fertile for only twelve to twenty-four hours, it is usually fertilized in the fallopian tube.

Uterus

The *uterus* (womb) is a hollow, pear-shaped organ (Figure 10.2). Located in the pelvis, it is slightly above and behind the bladder, but in front of the rectum. About three inches long and two inches wide, its walls are thick and very muscular. In pregnancy it may stretch to over twelve inches in length as it adapts itself to the

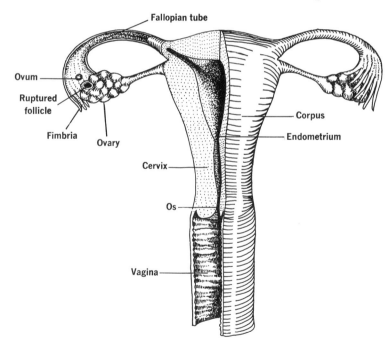

Figure 10.3 Uterus and related organs, front view. Notice release of ovum from right ovary.

growing fetus. Its necklike portion, or *cervix*, extends downward into the *vagina*.

The uterus is loosely suspended in position by several ligaments and is normally tilted forward slightly. A loosening of these ligaments as a result of childbearing may cause it to be tilted backwards.

Its inner layer, or *endometrium*, is richly supplied with blood vessels and glands. Following ovulation, the egg descends through the fallopian tubes and enters the cavity of the uterus. If it has been fertilized, it becomes embedded within the endometrium within three to four days.

Vagina

The *vagina* is a muscular tube three to four inches long. It serves as the female organ for intercourse and also as the birth canal at childbirth. In young girls the external opening of the vagina is partially closed by a membrane called the *hymen*. Varying in size and thickness, the hymen may remain intact until the first intercourse. Since it may also be broken or reduced by active sports, its absence need not be taken as a sign of lack of virginity. The hymen may also be altered by the use of tampons (vaginal insertions used to absorb menstrual discharge) or by a physician as part of a medical examination. On the other hand, the hymen may be so well developed in some girls that a physician may need to cut it surgically in a premarital examination in order to reduce the discomfort which might otherwise accompany the first intercourse.

External Genitalia

Two pairs of liplike structures surround the external opening of the vagina; these are the *vulva*. The outer and larger pair is the *labia majora*; the inner, smaller pair is the *labia minora*. The external opening of the urethra lies just in front of the vaginal opening (Figure 10.2).

Above the urethral opening, at the point where the labia majora come together, is a small erectile organ called the *clitoris*. This sensitive organ is somewhat similar to the penis in the male, although much smaller in size, in that during sexual excitement it becomes erect. The clitoris is the chief site of sexual excitement and sexual satisfaction in the female.

MENSTRUAL CYCLE

As we have already noted, the ovaries serve two functions: the production of eggs (ova), and the secretion of the female sex hormones. Once a month a mature follicle ruptures and releases an egg.

Menstruation

As the egg is maturing, the mature follicle secretes a hormone which causes the capillaries in the endometrium lining the uterus to become engorged with blood. Glands, also present in the lining, grow and become twisted; the lining cells of the glands secrete a fluid that will nourish the fertilized ovum. This action occurs as a preparation for the arrival of a fertilized egg. If the egg is not fertilized, about two-thirds of the inner lining will slough away in a process called *menstruation*. The menstrual products discharged are composed of the sloughed lining and blood. The beginning of menstruation marks the first day of the *menstrual cycle*. The length of time this discharge lasts is called the *menstrual period*. A period will usually last three to five days, although for some women periods from two to eight days can be considered normal. The amount of menstrual flow may vary somewhat from month to month and is greater in volume for some women than for others. The amount of blood lost usually ranges from two to four tablespoonfuls (25-60 ml.) per menstruation.

An average menstrual cycle is about twenty-eight days. But a woman with an average twenty-eight day cycle may, on occasion, range between twenty-one and thirty-two days. Some women are known to have normal cycles as short as twenty days and as long as forty-five days. The sequence of events of the menstrual cycle are shown in Figure 10.4.

Menstruation usually begins about the age of twelve. This beginning of menstruation is referred to as *menarche*. Menstruation continues until a woman is between forty and fifty years of age, when the ovaries become less active. Eventually menstruation ceases entirely; this phase is referred to as *menopause*.

Disturbances of Menstruation

Variations in the normal course of menstruation and the menstrual cycle are common. Here are four examples of variations:

1. Some women bleed excessively during the monthly period, a condition known as *menorrhagia*. It may be caused by disturbances in the amounts of reproductive hormones produced or by abnormal conditions within the uterus.

2. Bleeding may occur at odd times during the cycle. Such bleeding may be slight and fleeting or persistent throughout much of the cycle. Such "spotting" may be due to hormonal disturbances, inflammation of the endometrium, or disturbances of the muscles of the uterus. Unsuspected tumors may also be a cause of such irregular bleeding.

3. An entire menstruation may be missed (*amenorrhea*). The most common reason for a missed menstruation is pregnancy, but a missed menstruation by itself is not a positive sign of pregnancy.

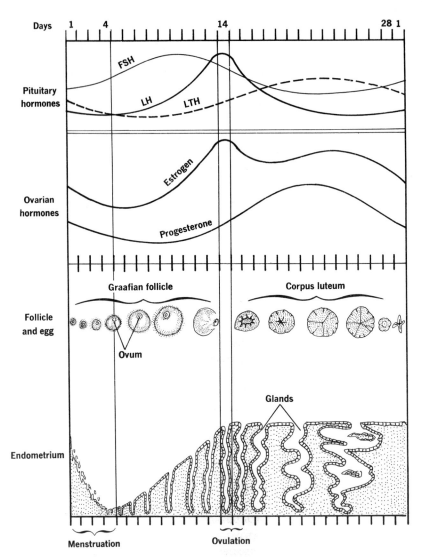

Figure 10.4 Diagram comparing events of typical menstrual cycle with days of month. Top to bottom: pituitary hormones, ovarian hormones, egg and follicle, thickness of endometrium and development of glands.

Amenorrhea may also be caused by inadequate amounts of reproductive hormones, either temporarily or over a long period of time. If menstruation never begins in a girl, it may indicate that ovaries are absent or have never developed. In such cases, the girl would show other signs of underdevelopment, such as immature breasts, since their development depends on hormones that must come from developed ovaries.

4. *Dysmenorrhea* (painful and difficult menstruation) may be experienced. This may include cramps, low abdominal pain, aching of the thighs and back, headache, nausea, and perhaps emotional changes. The causes of dysmenorrhea are not fully understood, and treatment commonly involves the use of pain-deadening drugs.

Menstruation will permanently stop in the event a woman has a *hysterectomy*. This is an operation that removes the uterus. It may be performed to alleviate benign tumors or bleeding conditions, complications arising as a result of childbirth, or cancer. Such an operation not only terminates further menstruation, but also terminates the possibility of bearing children.

Menstrual irregularities that are out of the ordinary may be due to a complication of pregnancy, a developing disease, or insignificant temporary causes. Such irregularities are reasons to see a physician promptly.

Menopause

Menstruation continues until a woman is between forty and fifty years of age, when the cycles become less frequent and the amount of bleeding decreases and then ceases. Menopause has been known to occur in women as young as thirty-six years of age, and in others as old as sixty years. Occurring gradually, these events are often called the "change of life" or the *climacteric period*. The woman's internal sex organs shrink, she no longer produces eggs, and she cannot become pregnant.

Although natural in a woman, the climacteric period is often accompanied by some disturbances. Some women experience "hot flashes," anxiety and depression, irritability, and headaches. These feelings are the result of hormonal changes. Once the change is completed, such symptoms should disappear.

In the event symptoms are severe, estrogenic hormones are frequently given to replace the natural estrogen which is declining too rapidly. Such treatment will relieve the symptoms.

Reproductive Hormones

The menstrual cycle is under the master control of the *anterior pituitary gland*, located at the base of the brain. This gland secretes, among others, three hormones (the gonadotropic hormones) which control the ovaries. Pituitary gonadotropins and their effects are shown in Table 10.1.

Table 10.1 Pituitary Gonadotropins

HORMONE	EFFECT
Follicle-stimulating hormone (FSH)	FSH directs the development and activity of the graafian follicles. It causes the follicle to secrete estrogen.
Luteinizing hormone (LH)	LH works with FSH to stimulate continued estrogen production. It triggers ovulation, thereby initiating formation of the corpus luteum and causing it to secrete both estrogen and progesterone.
Luteotropin (LTH)	LTH, also called prolactin, helps to prolong estrogen and progesterone production by the corpus luteum. It causes milk secretion by the mammary glands after the birth of the baby.

Under the influence of the gonadotropic hormones, the ovaries become active. In turn, the ovaries, serving as endocrine glands, secrete several hormones essential to the menstrual cycle. These ovarian hormones and their effects are summarized in Table 10.2.

Table 10.2 Ovarian Hormones

HORMONE	SOURCE	FUNCTION
Estrogens	Ovary (egg follicle); placenta during pregnancy	Growth of the female sexual organs and promotion of secondary sexual characteristics of the female, such as breasts, hair distribution, voice, and bone structure. Growth of the endometrium; inhibits production of FSH, increases production of LH.
Progesterone	Ovary (corpus luteum, the ruptured egg follicle); placenta during pregnancy	Primary function is to maintain endometrium of uterus for fertilized egg. Causes swelling of breasts, but not milk production. Causes salt and water retention in body. Inhibits production of LH.

FOLLICLE-STIMULATING HORMONE

A follicle-stimulating hormone (FSH) is produced by the pituitary gland and carried to the ovary by the bloodstream. In the ovary it stimulates a follicle to greatly enlarge and mature. This enlarging follicle secretes quantities of *estrogen*.

ESTROGEN

The increased presence of estrogen causes several things to happen. It influences the development of secondary sex characteristics—the development of the vagina and the uterus, pubic hair, and enlarging of the breasts, the pleasing feminine proportions, and the female sexual drive. Estrogen stimulates the development of the endometrium, helping to make it ready to receive the released egg, which may be fertilized. It is carried by the bloodstream to the pituitary gland, which then slows down its FSH production. It also stimulates the pituitary gland to produce a *luteinizing hormone* called LH.

LUTEINIZING HORMONE (LH)

LH controls the activities of the *corpus luteum*. After the follicle of the ovary has released its egg, the remains of the follicle contained in the ovary are called the corpus luteum (which means "yellow body").

PROGESTERONE

The corpus luteum has the important task of producing *progesterone*. This hormone, along with estrogen, *matures* the endometrium in preparation for receiving a fertilized ovum. It also keeps the endometrium in a healthy condition preventing menstruation. If there is not enough progesterone available, the lining inside the uterus begins to slough away. Then, if the woman becomes pregnant, she will suffer a miscarriage. Thus the LH has the important task of protecting the corpus luteum so that it can produce enough progesterone to keep the endometrium in good condition.

LUTEOTROPIN (LTH)

Keeping the corpus luteum in good health is so important that the pituitary gland produces a hormone to help the LH. Called *luteotropin*, or LTH, this hormone works with LH in stimulating the corpus luteum to produce progesterone. LTH is also important in a mother's production of milk in the breasts. LTH is, in fact, sometimes called *prolactin*, referring to milk.

Not all of the progesterone is used by the uterus lining. Some of it, when carried by the blood stream back to the pituitary gland, will cause the gland to reduce its production of both LH and LTH. When this reduction occurs, the corpus luteum will degenerate and be unable to produce enough progesterone. Then, the lining of the uterus cannot be maintained, and it begins to disintegrate. This disintegration is menstruation.

About this time the FSH production increases again, and, following the three to five days of menstruation, the lining inside the uterus is once again restored. The woman is then in her next menstrual cycle.

In the event of pregnancy, special tissues (the *placenta*) that form around the embryo give off LH and LTH (just as the pituitary gland does) to maintain the corpus luteum and progesterone production so that menstruation will not occur. If menstruation *does* occur, the young embryo is lost and the pregnancy ends. (Some women may spot after pregnancy begins, but do not actually menstruate.)

The interaction of the hormones that regulate the menstrual cycle is summarized in Figure 10.4. Toward the end of each menstrual cycle, the lowering of the amounts of ovarian hormones, progesterone and estrogen, allows the pituitary gland to again increase the amount of FSH in the blood. This marks the beginning of a new menstrual cycle.

The following is a summary description of the menstrual cycle:
1. At puberty the *pituitary gland* starts giving off FSH which affects the ovary. This hormone causes a new follicle to begin developing about once every twenty-eight days.
2. The growing follicle (in the ovary) produces estrogen which (a) prepares the uterus and its lining, (b) slows down the FSH production in the pituitary gland, and (c) speeds up the LH production in the pituitary gland.
3. FSH and LH both help ovaries to produce and release an egg (ovulation).
4. After ovulation, LH causes the ruptured follicle to become a *corpus luteum*.
5. The corpus luteum (in the ovary) produces progesterone, which (a) keeps the uterus and its lining in good condition for a possible fertilized egg, and (b) slows down the LH production in the pituitary gland.
6. An LH reduction in the pituitary gland causes the corpus luteum to "dry up" and stop producing progesterone.
7. A shortage of progesterone causes the lining of the uterus to give way, and menstruation begins.
8. A new follicle begins development.

PREGNANCY

Reproduction requires the union of a sperm with an egg. In order for this union to occur, sperm must be introduced into the vagina of the woman. This introduction is accomplished through *sexual intercourse.*

Once introduced into the vaginal canal, the sperm cells quickly move upward through the canal into the uterus and outward through the fallopian tubes toward the ovaries. Upon ovulation, the sperm surround the egg and one will penetrate (fertilize) it. Although only one sperm actually penetrates the egg, others help to detach (by enzymatic action) a covering over the egg. Eggs have been recovered from women's bodies with up to sixty sperm surrounding them.

The nine calendar months of pregnancy are commonly divided into thirds, or *trimesters,* of three months each. If an embryo is lost during the first or second trimesters (before it can survive) the loss is called an *abortion* or a *"miscarriage."* If a living child is born during the third trimester, but before full term, his birth is called a *premature birth*; if he is dead, it is known as a *stillbirth.*

It is often hard to estimate when a baby will be born because the exact day of fertilization is seldom known. In addition, not all pregnancies are of exactly the same length. But a rough estimate of the delivery date can be made by figuring 280 days (forty weeks) from the *beginning of the last menstrual period before pregnancy began.*

SUMMARY
   I. The Male Reproductive System
      A. Production of sperm
         1. Testes
            a. Produce sperm and male hormones.
            b. Carried in scrotum to lower temperature several degrees.
            c. Each testis composed of numerous seminiferous tubules, in which sperm are produced.
         2. Epididymis is the coiled tube on back of each testis in which sperm are held.
         3. Seminal ducts (vas deferens) carry sperm from each epididymis into the abdomen and empty into urethra.
      B. Fluid-producing glands
         1. Seminal vesicles—one empties into each seminal duct near the upper end.
         2. Prostate is just below urinary bladder, surrounding point where seminal ducts join urethra.
         3. Cowper's glands are two small glands ducted into the urethra.
         4. Secretions from above glands, plus sperm, is semen.
      C. Penis
         1. Erectile tissue becomes engorged with blood to produce erection.

2. End is called the glans.
3. Inability to attain erection is impotence.
4. Ejaculation is the expulsion of semen from the penis.
D. Male reproductive hormones
1. Principal male hormone is testosterone.
2. Production of testosterone is stimulated by hormone ICSH from pituitary.
3. Testosterone is produced by interstitial cells located between seminiferous tubules.
4. The hormone FSH from the pituitary stimulates sperm production.

II. The Female Reproductive System
A. Ovaries
1. Ovaries are the female gonads, producing eggs and hormones.
2. Eggs are enclosed in follicles within ovaries.
B. Fallopian tubes (oviducts)
1. Carry eggs from ovaries to uterus.
2. Have inner lining of cilia to carry eggs along.
3. Expulsion of egg from ovary is ovulation.
4. Fertilization of egg usually takes place in the fallopian tube, within twelve to twenty-four hours after ovulation.
C. Uterus
1. Contains growing fetus during pregnancy.
2. Necklike portion is cervix.
3. Inner lining is endometrium.
D. Vagina
1. Muscular tube three to four inches long.
2. Serves as organ of intercourse and as birth canal.
E. External genitalia
1. Liplike structures surrounding the opening of the vagina are called vulva or labia.
   a. Labia majora—outer pair.
   b. Labia minora—inner pair.
2. Clitoris is small erectile organ at point where labia majora join in front of vaginal opening.

III. Menstrual Cycle
A. Menstruation
1. Sloughing away of the endometrium.
2. Lasts an average of three to five days.
3. Average cycle is twenty-eight days, though longer or shorter cycles may be normal.
4. First menstruation is called menarche.
5. Cessation of menstrual cycles is menopause.

B. Disturbances of menstruation
   1. Menorrhagia or excessive bleeding.
   2. Irregular bleeding or "spotting."
   3. Amenorrhea or a missed menstruation.
   4. Dysmenorrhea is painful or difficult menstruation, which may result in various physical or emotional ailments.
C. Menopause
   1. Cessation of menstruation.
   2. Occurs as part of the climacteric.
   3. May be accompanied by hot flashes, anxiety, and depression, irritability, and headaches.
   4. Estrogenic hormones may be given to relieve adverse effects of menopause.
D. Reproductive hormones
   1. Menstrual cycle controlled by anterior pituitary gland, located at base of brain.
   2. Follicle stimulating hormone (FSH) from pituitary stimulates follicle in ovary to mature and to secrete estrogen.
   3. Estrogen
      a. Stimulates development of female characteristics.
      b. Stimulates development of endometrium.
      c. Inhibits FSH production by pituitary gland.
      d. Stimulates pituitary gland to produce LH.
   4. Luteinizing hormone (LH) stimulates ovulation, formation of corpus luteum from empty follicle, and production of progesterone by corpus luteum.
   5. Progesterone
      a. Matures the endometrium.
      b. Prevents menstruation.
   6. Luteotropin (LTH) (prolactin)
      a. Produced by pituitary gland.
      b. Stimulates the corpus luteum to produce progesterone.
      c. Stimulates milk production.

IV. Pregnancy
   A. Many sperm must surround egg to detach a covering by enzyme action, so that one can penetrate and fertilize.
   B. Delivery date can be estimated by figuring forty weeks from the beginning of the last menstrual period before conception.

QUESTIONS FOR REVIEW
1. What are the male and female gonads?
2. Why are the testes carried in the scrotum?
3. What are the sources of seminal fluid?

4. What is an erection of the penis?
5. What are the effects of testosterone?
6. What carries eggs down the fallopian tubes?
7. What is the menstrual flow?
8. Define menarche and menopause.
9. Give the source and functions of each of the following hormones in the female:
   a. FSH
   b. LH
   c. LTH
   d. Estrogen
   e. Progesterone

Chapter 11
ENDOCRINE GLANDS AND HORMONES

As we have seen, the human body is a very complicated structure. Yet, we all expect our bodies to respond immediately and smoothly to our express commands. We further expect our bodies to automatically maintain certain essential procedures (such as breathing and heartbeat) around the clock. All of this complex organization is coordinated and regulated by certain parts of the nervous system and by certain chemical substances carried in the blood. This chapter will discuss the chemical regulators carried in the blood.

GLANDS AND HORMONES

A gland is an organ that takes materials from the body, makes new substances from them, and secretes them. The substance produced is either used *elsewhere* in the body (as with the pituitary gland) or is eliminated from the body (as with a sweat gland).

The body contains various kinds of glands, which may be grouped into the following two main divisions: 1. *Exocrine*, or *ducted*, glands that give off their secretions through a duct into a cavity or onto the surface of the body, as with sweat glands or salivary glands; and 2. *Endocrine*, or *ductless*, glands that give off their secretions into the blood or body fluids. Some glands are clearly one or the other; while others, such as the pancreas, combine both types of action. The secretions given off by endocrine glands are known as *hormones*.

Hormones from the endocrine glands produce striking effects on specific tissues and, like vitamins, are needed in relatively small amounts to produce their effect. The tasks these hormones ("chemical messengers") are called upon to do are quite varied. They may stimulate exocrine glands to produce secretions; stimulate endocrine glands to action; affect growth and development, or personality; regulate chemical reactions within general body cells; regulate contraction of muscles; or control sex processes. The action of each hormone is *specific* on a given tissue.

THE ENDOCRINE GLANDS

The endocrine glands are quite isolated from each other in the body (Figure 11.1). The endocrine glands of the body include the: (1) pituitary (hypophysis), (2) thyroid, (3) parathyroids, (4) adrenals, (5) pancreas, and (6) gonads. The structure and function of each of these will be discussed. Not included are the pineal and thymus glands. The pineal gland secretes a hormone called melatonin. Its effect on humans is still uncertain. The thymus gland (considered by some as an endocrine gland) plays a major role in setting up lymphocyte-production of the lymph nodes, thus helping the development of antibodies.

The Pituitary (Hypophysis)

The pituitary is suspended from the underside of the brain, and is protected by a small depression in the base of the skull. Investigators once thought the gland helped in the production of phlegm (mucus) and gave it the name *pituitary* (which means "phlegm"). A more preferable synonym today is *hypophysis*, which means "undergrowth." This chapter will refer to it as the pituitary simply because this is commonly understood.

Less than one-half inch in diameter, it is divided into three lobes: anterior (front), intermediate, and posterior (hind).

ENDOCRINE GLANDS AND HORMONES   141

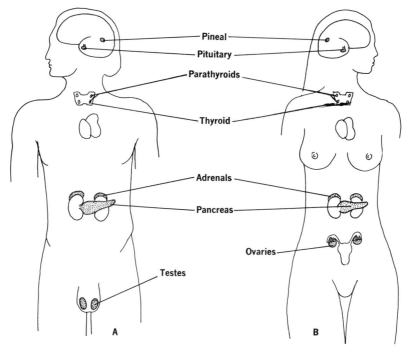

Figure 11.1 Location of endocrine glands. (A) male. (B) female.

ANTERIOR LOBE

The anterior lobe comprises the largest portion of the gland. Some nerve fibers connect parts of it to the brain. Actions by the central nervous system can affect its secretions. The hormones of the anterior lobe (called *tropins*) help to maintain other endocrine glands and stimulate them to produce their own hormones. Six hormones are known to be secreted by the anterior lobe (Figure 11.2).

1. *Growth hormone* (GH). Also called somatotropic hormone (SH), it causes growth of all tissues of the body that are capable of growing, both by increased cell division and by increased size of the cells. Excess of GH in early life causes *giantism*; excess in adults causes *agromegaly* in which the jaws, bones, hands, and feet become enlarged and features become coarse. An insufficient supply in adolescence causes *dwarfism*.

2. *Gonadotropic hormones.* These hormones affect the sex glands (gonads) in both sexes and include:

a. *Follicle-stimulating hormone (FSH).* This causes the growth and maturing of follicle cells in the ovary of the female and causes them to secrete estrogens. It stimulates the development of sperm in the male.

142 THE HUMAN BODY

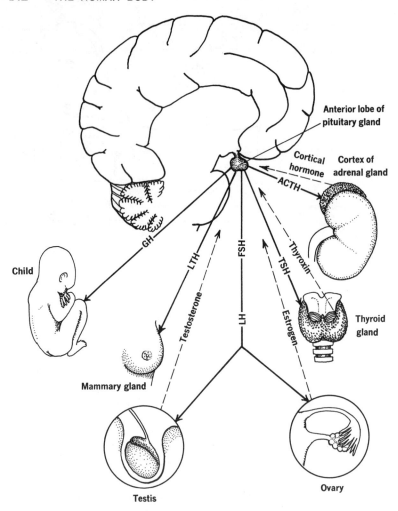

Figure 11.2 Location of pituitary gland and some hormones it produces. Note that some glands affected by its hormones in turn produce hormones which affect the pituitary gland.

    b. *Luteotropin (LTH),* or prolactin. This stimulates the corpus luteum in the ovary to produce progesterone and estrogens. It also stimulates the enlargement of mammary glands and the secretion of milk.

    c. *Luteinizing hormone (LH).* This helps the ovary in the development of follicle cells, the corpus luteum, and production of progesterone. It stimulates the interstitial cells in the testes of the male to produce testosterone (which is responsible for making the male look masculine) and is thus called the *interstitial-cell-stimulating hormone* (ICSH).

3. *Thyroid-stimulating hormone* (TSH). Also called the thyrotropic hormone, this affects the structure of the thyroid gland and stimulates its production of hormones.

4. *Adrenocorticotropic hormone* (ACTH). Also called corticotropin, this stimulates hormone production in the cortex of the adrenal gland.

INTERMEDIATE LOBE

The intermediate lobe gives off *melanocyte-stimulating hormone* (MSH) which has to do with the pigmentation or darkening of the skin.

POSTERIOR LOBE

The posterior lobe is known to secrete at least two important substances.

1. *Oxytocin.* This hormone stimulates the contraction of smooth muscles. During the time of childbirth, it causes the contraction of the uterus. Sexual stimulation of the female during intercourse increases the secretion of oxytocin, causing uterine contractions at this time. Oxytocin plays an important role during the process of lactation (milk production). At the time of milk production in the mother, it causes milk to be discharged into the ducts of the breasts so the baby can obtain it by sucking.

Oxytocin injections may be used in childbirth to prevent hemorrhage resulting from an overly relaxed uterus after delivery.

2. *Antidiuretic hormone* (ADH). Also called vasopressin, ADH promotes the reabsorption of water by the tubules of the kidneys. In the discussion on urine in Chapter 9, diuretics were described as substances that increase the flow of urine. An antidiuretic is a substance that would reduce the flow. ADH also serves to elevate blood pressure by constricting (reducing) the size of small arteries near the surface of the body. It may be used in cases of surgical shock to elevate blood pressure.

The Thyroid

The thyroid gland is a shield-shaped organ located in the neck near the junction of the larynx and trachea (Figure 11.1), or just below the "adam's apple." The gland consists of two lateral (side) lobes that lie on either side of the trachea, connected by an isthmus (bridge). Larger in the female than in the male, its size and shape will fluctuate with age, reproductive state, diet, time of the year, and a person's geographical location. It normally weighs less than one ounce.

As with the pituitary gland, its blood supply is exceptionally rich (probably more blood flows through this gland in proportion to its size than any other organ of the body). It is estimated that four to five quarts of blood flow through the gland every hour!

When the thyroid gland is removed from the body, there is a great reduction in the basal metabolic rate (BMR). Metabolism is the sum of all the chemical processes within the body. The basal metabolic rate is the smallest amount of energy required to maintain the functioning of the body. Thus it appears that the primary function of this gland is the regulation of the body's metabolic rate. This gland produces three hormones: *thyroxine, triiodothyronine,* and *diiodothyronine*. The function of all three is essentially the same; they differ only in speed and intensity of action. The specific action of thyroxine will be discussed.

Iodine is taken from the body by the cells of the thyroid gland. When the iodine-containing hormones are released from the gland, they combine with the proteins in the blood to form protein-bound iodine (PBI). For a physician to determine the activity of the thyroid gland all he needs to do is take a blood sample and measure the PBI. This gives him a clue to the person's rate of basal metabolism.

The roles of thyroxine in the body include:

1. Regulation of metabolic rates in the tissue cells of the body (it increases the rates that glucose is taken from the intestine and used by the body cells).

2. It stimulates growth and differentiation of tissues.

3. It influences conversion of glycogen (stored starch in the body) to glucose, thus raising the blood sugar level.

4. It influences both physical and mental development in the child and stimulates mental processes among all ages.

5. It increases heart rate.

The body can be greatly affected by either oversecretion or undersecretion of this hormone. Insufficient iodine in the diet can cause an enlargement of the gland, or simple *goiter*. An undersecretion early in life, causes *cretinism*, a condition marked by retarded skeletal growth, slowed mental development, large head, weak muscles, and slow speech. An undersecretion in an adult causes *myxedema*, which is marked by slow speech and body movements, rough skin, hair loss, puffy hands and face, and low metabolic rate. Any type of undersecretion may be termed *hypothyroidism*. Oversecretion, or *hyperthyroidism,* is characterized by protruding eyeballs (exophthalmic goiter), increased heart action, elevated body temperature, nervousness, sleeplessness, excess appetite, and weight loss.

The Parathyroids

The parathyroids are very small glands (under one-quarter

inch long) lying on the surface of the thyroid gland. Their number can vary from two to twelve, but the average number is four.

The parathyroids produce a hormone (*parathormone* or PTH) that plays a vital role in the metabolism of calcium and phosphorus. PTH maintains the proper calcium-phosphorus ratio in the blood and tissues by influencing calcium's absorption from the intestine, its deposition in the bones, mobilization from bones, and excretion by the kidneys.

Hypoparathyroidism, or undersecretion, due either to disease or gland removal, may cause muscle weakness, tetany, and irritability. *Tetany* is a condition characterized by highly increased sensitivity to external stimuli, from which painful spasms of the muscles occur. *Hyperparathyroidism*, or oversecretion, may cause kidney stones to form and the bones to be decalcified. Such bones may become soft and fracture easily. As with hormone imbalances from other glands, these abnormalities can usually be corrected medically.

The Adrenals

The two *adrenal glands*, sometimes referred to as the *suprarenal glands*, are perched atop the kidneys (Figure 11.1). Either name is self-descriptive. *Suprarenal* means "above kidney," and *adrenal* means "near kidney."

Each gland is pyramid-shaped and one to two inches across. The glands have a profuse supply of blood. Each one consists of two distinct parts: a central portion, the *medulla*, and an outer layer, the *cortex*. Each part secretes hormones distinct in chemical composition and in function. There is no relationship between the two parts other than location. The nerve supply of the medulla is more profuse than that of any other tissue.

THE CORTEX

The adrenal cortex is under the influence of ACTH from the pituitary gland, and is known to secrete over thirty different hormones, several of which are essential to the body. The removal of both adrenal cortices can cause death within three days. The essential cortical hormones are called *corticoids*, and are chemically similar to sex hormones.

These hormones have a number of effects on the body. They alter kidney function, carbohydrate metabolism, and circulation. They increase fat usage, help the body to resist infection, increase blood sugar, and help regulate the body's use of water and certain minerals. Some cortical secretions have an important relationship to other endocrine glands, such as the gonads.

A material extracted from cortical hormones is called *corti-*

sone, and is used in the treatment of certain diseases. It has been successful in relieving symptoms of rheumatoid arthritis and helps in the recovery of rheumatic fever patients.

Insufficient cortical production causes *Addison's disease.* This is characterized by weight loss, weakness, a bronzing coloration of the skin, and, in extreme cases, death. Excessive cortical production, called *Cushing's disease,* may cause unusual sexual development. Overactivity before puberty in boys may cause boys as young as five or six years old to develop beards and body hair and become sexually mature. In young girls it may cause premature menstruation, early breast development, and sexual drive. Oversecretion in older woman may cause sexual reversal—beard development, a low voice, and reduction in breast size. In men there may be breast development, and a change in hair distribution.

THE ADRENAL MEDULLA

The adrenal medulla is under the control of the sympathetic division of the nervous system and secretes two hormones, *epinephrine* (adrenalin) and *norepinephrine* (noradrenalin).

Under certain emergency situations (such as injury, excessive muscular exercise, infection, hemorrhage, cold, fever, burn, nervous shock, and lack of oxygen) epinephrine is released into the blood. The following are some of the effects of epinephrine on the body:
1. A more rapid and forceful heartbeat.
2. A greater flow of blood to the muscles, central nervous system, and the heart.
3. An increased output of glucose from the liver.
4. A rise in general blood pressure.
5. Decrease in the coagulation time of the blood.
6. Discharge of red blood cells from the spleen.
7. Increase in the depth and rate of respiration.
8. Contraction of smooth and skeletal muscles.

These things occur when a person is placed under emotional stress, such as when a person becomes frightened or angry. It is not idle counsel to admonish a friend during a spell of anger to "not get his blood pressure up." Such stress may cause a person to "turn white" in fear; to perform unusual physical feats in emergencies; to get "gooseflesh" in certain emotional situations; or to have his hair "stand on end" (more easily observed in some dogs than in humans).

By looking at these effects, it becomes easier to understand why epinephrine has certain uses in medicine. It has been used to relieve spasms of the air ducts (as in asthma) and make breathing easier; to constrict the small arteries of the mucous membranes and of the skin, thus reducing blood loss in minor operations; and to constrict blood vessels near the skin, thus prolonging the effect of anesthetics used on the skin.

Norepinephrine also tends to rally the body to situations of stress or emergency, in a similar manner.

The Pancreas

As already seen in our study of digestion, the pancreas generally lies directly below the stomach. Most of the gland is concerned with producing pancreatic enzymes for digestive activities, but within the gland are small scattered groups of cells called *islets of Langerhans*. These cells produce two hormones, *insulin* and *glucagon*.

Insulin promotes the use of glucose in body cells and thereby decreases blood sugar concentration. In the liver and muscles it increases the conversion of glucose (blood sugar) to glycogen (starch). Glucagon has the opposite effect—the conversion of glycogen to blood sugar in the liver.

Too much insulin in the blood, or *hyperinsulinism*, causes a rapid fall in the blood sugar level producing a condition known as *hypoglycemia*. One of the first symptoms of hypoglycemia may be blurred vision and an inability to focus the eyes. A feeling of drowsiness and yawning will usually occur, and the person may become excited or perspire or appear to be under the influence of alcohol. Convulsive seizures may occur, and finally the person may go into a coma and die.

Too little insulin, or *hypoinsulinism*, causes a rapid rise in the level of blood sugar, producing a condition known as *hyperglycemia*. This results in a series of symptoms known as *diabetes mellitus*. Excess sugar in the blood will cause the kidneys to be unable to reabsorb all of the glucose out of the tubule fluid. This means sugar will be found in the urine—the condition known as *glycosuria*. Sugar in the urine leads to further loss of water, and this leads to dehydration. The administration of insulin to diabetic patients rapidly restores the correct blood sugar level.

The Gonads

The gonads are endocrine glands of considerable importance due to their secretion of testosterone by the testes and of estrogen and progesterone by the ovaries. For the structure and function of the gonads the student is referred to Chapter 10.

SUMMARY
  I. Glands and Hormones
     A. Glands
        1. Exocrine (ducted)—secretions carried off through a duct.

148 THE HUMAN BODY

2. Endocrine (ductless)—secretions released into the blood.
3. Some glands combine both types of action.

B. Hormones
1. Products of endocrine glands.
2. Have specific actions on specific tissues.

II. The Endocrine Glands

A. The pituitary (hypophysis)
1. Located on underside of brain.
2. Has 3 lobes:
   a. Anterior lobe—hormones, called tropins, mainly stimulate other endocrine glands.
      (1) Growth hormone (GH), also called somatotropic hormone (SH), stimulates growth of body.
      (2) Gonadotropic hormones stimulate gonads in both sexes.
         (a) Follicle-stimulating hormone (FSH) — causes growth of egg follicles in ovary and production of estrogen in female; stimulates sperm production in male.
         (b) Luteotropin (LTH), or prolactin stimulates corpus luteum to produce progesterone and estrogens; stimulates mammary glands (breasts).
         (c) Luteinizing hormone (LH) stimulates formation of corpus luteum and production of progesterone by ovary. In male it is called interstitial-cell-stimulating hormone (ICSH) and stimulates testosterone production.
      (3) Thyroid-stimulating hormone (TSH), also called thyrotropic hormone stimulates production of hormones by thyroid gland.
      (4) Adrenocorticotropic hormone (ACTH) stimulates hormone production in cortex of adrenal gland.
   b. Intermediate lobe releases melanocyte-stimulating hormone.
   c. Posterior lobe releases at least two hormones.
      (1) Oxytocin stimulates contraction of smooth muscles, especially of uterus.
      (2) Antidiuretic hormone (ADH), or vasopressin, promotes reabsorption of water by kidney tubules; raises blood pressure by contracting small arteries.

B. The thyroid
1. Located in neck.
2. Hormones include thyroxine, triiodothyronine, and diiodothyronine.

3. All have the same functions:
   a. Stimulate basal metabolic rate.
   b. Stimulate growth and differentiation of tissues.
   c. Raise blood sugar level.
   d. Stimulate physical and mental development in children and stimulate mental processes among all ages.
   e. Increase heartbeat.
4. Undersecretion is hypothyroidism; oversecretion is hyperthyroidism.

C. The parathyroids
   1. Several (average is four) very small glands lying on surface of thyroid gland.
   2. Hormone is parathormone (PTH)—important in metabolism of calcium and phosphorus.

D. The adrenals
   1. One sits as cap on each kidney.
   2. Each has two distinct parts.
      a. The cortex
         (1) Secretes over thirty hormones, called corticoids.
         (2) Corticoids have a variety of functions and are essential for life.
      b. The adrenal medulla
         (1) Under control of sympathetic nervous system.
         (2) Hormones are epinephrine (adrenalin) and norepinephrine (noradrenalin).
         (3) Both serve to prepare body for emergency action.

E. The pancreas
   1. Lies directly below the stomach.
   2. A dual gland, produces digestive enzymes (exocrine function) and hormones (endocrine function).
   3. Hormones are insulin and glucagon, both produced in small scattered groups of cells called islets of Langerhans.
      a. Insulin promotes use of glucose in body cells and lowers blood sugar level.
      b. Glucagon has opposite effect.
      c. Hypoinsulinism causes diabetes mellitus.

F. The gonads
   1. Important endocrine glands.
   2. See chapter on reproduction for discussion.

QUESTIONS FOR REVIEW

1. Contrast exocrine and endocrine glands.
2. What are tropic hormones or tropins?

3. What is a PBI test? What is its significance?
4. What is cretinism?
5. Which part of the adrenal gland is less essential for life?
6. Where are the islets of Langerhans?

Glossary

abdomen	The part of the body extending from the diaphragm above to the pelvis below.
abscess	A pus-filled cavity, such as a boil.
acromegaly	A condition resulting when production of growth hormone resumes in an adult.
aerobic	Metabolizing only in the presence of molecular oxygen.
afferent	Bearing or conducting inward.
albuminuria	Presence of albumin (protein) in the urine.

alveolus	A small cavity; an air sac which is the terminal dilatation of the bronchioles in the lungs; a tooth socket.
artery	A vessel carrying blood away from the heart.
astigmatism	A focusing defect of the eye in which light rays fail to focus on the same plane, thus forming indistinct images.
bolus	A rounded mass of soft material.
bursa	A closed sac lined with synovial membrane containing fluid, found over an exposed and prominent part of a bone.
calorie	A unit of heat. The large or kilocalorie (C) is the amount of heat required to raise 1 Kg. of water from 15° to 16° C.
capillary	One of the microscopic vessels forming a network between the arteries and veins; through its walls materials enter and leave the blood.
carbuncle	Several boils occurring together as a mass.
cartilage	The gristle or elastic substance attached at the joints of bones; or forming certain parts of the body (ear, nose, etc.)
chyme	The semifluid mass of partly digested food passing from the stomach into the duodenum.
cortex	The outer, or surface layer of an organ.
deciduous	Not permanent; to fall off, or out, at maturity.
diastole	In each heartbeat, the period of dilation or relaxation of the heart muscle.
diuretic	A substance that increases urine flow.
efferent	Bearing or conducting outward.
ejaculation	The expulsion of semen by the male.
endocrine gland	A ductless gland that releases its product (a hormone) directly into the blood.
enzyme	A catalyst produced within a living organism which accelerates specific chemical reactions.
excretion	The elimination of wastes from the body; that which is eliminated.

GLOSSARY

exocrine gland	A gland that releases its secretion through a duct.
fetus	The unborn child after about eight weeks of development.
genetic	Inherited; pertaining to the inheritance of characteristics.
glaucoma	Excessive pressure within the eyeball.
glycosuria	Presence of glucose (sugar) in the urine.
gonadotropic	Having a gonad-stimulating effect.
gristle	See cartilage.
hematuria	Presence of blood in the urine.
hormone	A product of an endocrine gland; produced by the gland, carried by the blood, and acting at some other point in the body.
hyperopia	Farsightedness; light rays from close objects are focused toward some point behind the retina.
impotence	Inability to attain or maintain erection of the penis.
intramuscular	Within a muscle.
keratin	A very tough, insoluble protein which is the principle constituent of epidermis, hair, and nails.
lateral	Toward the side; away from the median plane.
medulla	The innermost or middle part of an organ.
melanin	A yellow to black pigment produced in the epidermis by special cells called melanocytes.
menarche	The beginning of the menstrual cycles.
menopause	Cessation of menstrual cycles.
menstruation	The cyclic, uterine bleeding resulting from the breakdown of the endometrium of the uterus.
myopia	Nearsightedness; light rays from distant objects come to their focal point in front of the retina.

peptide	An intermediate product in protein digestion, consisting of two or more amino acids.
peptone	An intermediate product in protein digestion, consisting of two or more peptides.
polypeptide	An intermediate product in protein digestion, consisting of three or more amino acids.
posture	The attitude of the body.
postural	Pertaining to posture.
presbyopia	A visual defect in which the lens loses its elasticity, being unable to adjust the eye to varying distances.
protein	An organic compound consisting of a chain of many amino acids.
proteose	An intermediate product in protein digestion, between a protein and a peptone.
puberty	The beginning of sexual maturity.
reflex action	An automatic, unlearned response to a certain stimulus, carried out without conscious effort or thought.
septum	A dividing wall, as between two adjacent chambers of the heart.
sphygmomanometer	An instrument for measuring blood pressure.
strabismus	A condition resulting from muscle imbalance in the eye in which one eye may be turned in, out, up, or down.
systole	In each heartbeat, the period of contraction of the heart muscle.
theory	A formulated hypothesis supported by a large body of observation and experiments.
unconscious	Insensible; not receiving any external stimulus.
vein	A vessel carrying blood toward the heart.
vital organ	An organ necessary for the continuation of life.

Bibliography

Asimov, Issac, *The Human Brain: Its Capacities and Functions*, Boston, Houghton Mifflin, 1964.
Baker, A. B., *Clinical Neurology*, 2nd. ed., Vols. 1–4, New York, Harper & Row, 1962.
Brash, James C., ed., *D. J. Cunningham's Manual of Practical Anatomy*, 12th. ed., New York, Oxford University Press, 1958.
Eastman, Nicholson J., and Louis M. Hellman, *Williams Obstetrics*, 13th ed., New York, Appleton–Century–Crofts, 1966.
Grollman, Sigmund, *The Human Body*, 2nd. ed., New York, The Macmillan Co., 1969.
Guyton, Arthur C., *Textbook of Medical Physiology*, 3rd. ed., Philadelphia, W. B. Saunders Co., 1966.

Jones, Kenneth L., Louis W. Shainberg, and Curtis O. Byer, *Sex*, New York, Harper & Row, Publishers, 1969.

Kimber, Diana Clifford, Carolyn E. Gray, Caroline E. Stackpole, Lutie C. Leavell, Marjorie A. Miller, and Florence M. Chapin, *Anatomy and Physiology*, 15th. ed., New York, The Macmillan Co., 1966.

Langley, L. L., E. Cheraskin, and Ruth Sleeper, *Dynamic Anatomy and Physiology*, 2nd. ed., New York, McGraw-Hill Book Co., Inc., 1963.

Miller, J. J. and C. R. Wells, *Your Teeth: and how to keep them*, New York, Lantern Press, 1951.

Scientific American, *The Living Cell*, San Francisco, W. H. Freeman and Co., 1965.

Smith, Bernard H., *Principles of Clinical Neurology*, Chicago, Year Book Medical Publishers, 1965.

Strand, Fleur L., *Modern Physiology; The Chemical and Structural Basis of Function*, New York, The Macmillan Co., 1965.

Tuttle, W. W., and Bryon A. Schottelius, *Textbook of Physiology*, 16th. ed., St. Louis, The C. V. Mosby Co., 1969.

Index

Abdomen, 151
   muscles of, 24, 25
Abdominal muscles, in urination, 118
Abnormal behavior, *see* Neurological
     disorders
ABO system, *see* Blood groups
Abortion, 135
Abscesses, 151
     See also Boils
Absorption, of food, 95, 102
Acetylcholine, 34, 35
Achilles Tendon, 26
Acid, stomach, *see* Stomach acid
Acne, 64, 68
   treatment of, 68
Acromegaly, 141, 151
ACTH, *see* Adrenocorticotropic
     hormone
Acuity, *see* Vision
Adam's apple, 143
Addison's disease, 146
Adductor muscles, 26
Adenosine triphosphate (ATP), 18
ADH, *see* Antidiuretic hormone
Adipose, 114
Adrenal glands, 145–147
Adrenalin, *see* Epinephrine
Adrenocorticotropic hormone (ACTH),
   143
Aerobic, 151
Afferent, 151
Afferent arterioles, 116
Air sacs, 89
Air, *see* Lungs, capacity of
Albuminuria, 118, 151
Alcohol, absorption of, 104
   and urine, 117–118
Allergies, 70
All-or-none law, 17
   and cardiac muscle, 27

Alveolar socket, 98
Alveolus, 152
Alveoli, 89, 90
Amenorrhea, 129, 130
Amylopsin, *see* Pancreatic amylase
Anemia, 83
Anesthesia, 41
Anesthetics, 147
Anger, 40
Ankle, 6
Anterior abdominal muscles, 25
Anterior lobe, of pituitary gland, 141
Anterior pituitary gland, 131
Antibiotics, and boils, 65
Antibodies, 70
Antidiuretic hormone (ADH), 143
Antigen, 70
Anus, 105
Anvil (incus), 54
Anxiety, and menopause, 131
Aorta, 76
Aortic valve, 76
Apex, of tooth, 98
Appendicular skeleton, 1, 4
   parts of, 4
Appendix, 105
Aqueous humor, of eye, 46
Arm, muscles of, 24
     See also Forearm
Arterial branches, of aorta, 76
Artery, 152
Ascorbic acid, *see* Vitamin C
Asthma, 41
Astigmatism, 48, 50, 152
Athletes, 8
Athlete's foot, *see* Ringworm
ATP, *see* Adenosine triphosphate
Atrioventricular node (AV node), 77
Atrium, of heart, 75, 76, 77
Auditory canal, 53

158 INDEX

Auditory nerve, 54
Auricle, 53
Autonomic nervous system, see Nervous system
AV node, see Atrioventricular node
Axial skeleton, 1
   number of bones, 3
   parts of, 3
Axon, 33, 34

Baby teeth, see Temporary teeth
Back, muscles of, 20
Baldness, 71
Ball and socket joints, 9
Basal metabolic rate, 144
Behavior, abnormal, see Neurological disorders
Bicarbonate compounds, 92
Biceps brachii, 24
Biceps femoris, 26
Bicuspid teeth, 97
Bile, 104
Biological oxidation, 86, 87
Birth canal, see Vagina
Birthmarks, 69
Blackheads, see Acne
Blind spot, 47
Blindness, 45
Blindness in the United States, see Glaucoma
Blood, amount of in body, 80
   amount of through heart, 77
   antibodies, 82–83
   lost in menstruation, 129
   parts of, 80
   transfusions, 82–83
   in urine, 118
Blood clotting, 10, 81–82
Blood count, 82
Blood groups (ABO), 82–83
Blood pressure, 40, 41
   measurement of, 78
Blood sugar, see Glucose
Blood vessels, rupture of, 10
Blue baby, 80
Blurry vision, see Vision
BMR, see Basal metabolic rate
Body, movements of, 13
Body energy, 18
Body heat, 18
Body temperature, 61–62
   and sperm production, 122
   and sweat, 61
   regulation of, 61–62
Boils, 65
Bolus, 97, 152
Bone marrow, 10
   and red blood cells, 80
Bones, growth of, 6
   in fetus, 6
   long bones, 7
   movement of, 8, 14
      See also Joints, Muscles
   number of, 3
   shape and thickness of, 8

Bony callus, 10
Bowman's capsule, 115, 117
Brachial artery, 78
Brachialis, 24
Brain, parts of, 35
Brain stem, 35, 37
Breastbone, see Sternum
Breasts, 133
Breathing, 37, 39, 90
Bronchi, 89
Buccinator, 19
Bundle of His, 77
Bursa, 9, 152
Bursitis, 9
Buttock, see Hip and buttock

Calcification, 6
Calcium, 6
Calf, of leg, 26
Callus, 10
Calorie, 152
Calyx, 114
Capillary, 75, 76, 152
Carbohydrates, 18
   metabolism of, 104
Carbon dioxide, 18, 86
   as waste, 113
   production of, 92
   transport of, 92
Carbuncle, 152
      See also Boils
Cardiac muscle, 13, 75
   and cell regeneration, 27
   structure of, 27
      See also Heart
Carpals, 6
Cartilage, 1, 4, 152
Cartilage cells, 6, 7
Castration, 125
Cataract, 50, 51
Cecum, 105
      See also Large intestine
Cell body, 33
Cellular respiration, see Biological oxidation
Central nervous system, see Nervous system
Cerebellum, 35, 37
      See also Brain
Cerebral hemispheres, 36
Cerebro-spinal fluid, 38
Cerebrum, 35, 37
   lobes of, 36
      See also Brain
Change of life, 131
Charley horse, 25
Chemical messengers, see Hormones
Chemicals, and the eye, 52
Chest cage, 4
Chewing, see Mastication
Childbearing, 128
Cholinesterase, 35
Choroid, of eye, 47
Chromosomes, 123
Chyme, 100, 152

INDEX 159

Ciliary body, of eye, 46
Circulatory system, 74–83
    parts of, 74
Circumcision, 124
Clavicle (collar bone), 4, 20, 23
Cleanliness, and athlete's foot, 66
    and skin care, 63–64
Clitoris, 128
Coagulation, see Blood clotting
Cochlea, 54
Collar bone, see Clavicle
Colon, see Large intestine
Color blindness, 50
Color vision, see Cones
Complemental air, 91
Cones, 47, 48, 50
Conjunctiva, 48
Conjunctivitis, 50
Connective tissues (ligaments), 9
Contact lenses, 51
Convex lenses, 50
Cootie (body louse), see Lice
Cornea, 46, 50
Coronary vessels, 77
Corpus luteum, 133, 142
Corpuscles, 80
Cortical, oversecretion of, 146
Corticoids, 145
Corticotropin, see ACTH
Cortisone, 146
Cortex, 152
    of adrenal gland, 145
    of cerebellum, 37
    of cerebrum, 36
    of kidney, 114
Cosmetics, and complexion, 64
    and the eye, 52
Cowper's glands, 123
Crabs (pubic louse), see Lice
Cranial nerves, see Nerves, cranial
Cranium, 3
Cretinism, 144
Crown, of tooth, 97
Crystalline lens, 46
Cushing's disease, 146
Cuspid teeth, 97
Cuticle, 63
Cyanocobalamin (vitamin B_{12}),
    deficiency symptoms, 107
    function of, 107
    properties of, 107
    sources of, 107

Dandruff, 70–71
Deafness, 45
    conductive, 54
    perceptive, 54
Decarboxylation, 86
Deciduous, 152
Deep back muscles, 23
Defecation, 105
Deltoid muscle, 23
Dendrite, 33
Dentin, of tooth, 97
Deodorants, 64

Dermatologist, 71
Dermis, 60
    See also Skin
Diabetes mellitus, 118, 147
Diaphragm muscle, 23
    and breathing, 90
Diastole, 77, 152
Diet, and skin nourishment, 64
Diffusion, of gases, 88
Digestion, 39, 95
    processes of, 37
Digestive enzymes, 102
Digestive system, 95–108
    parts of, 96
    size of, 96
Diiodothyronine, 144
Disorders, of ear, 54–55
    of eye, 48–51
    of nervous system, 42
Distal convoluted tubule, 115
Diuretic, 117, 152
Dorsal root, 38
Ducted glands, see Exocrine glands
Ductless glands, see Endocrine glands
Duodenum, 100, 102
Dwarfism, 141
Dysmenorrhea, 131

Ear, care of, 55
    disorders of, 54–55
    parts of, 53–54
    structure of, 53–54
Ear doctor, see Otologist
Eardrum, see Tympanic membrane
Eczema, see Allergies
Efferent, 152
Eggs, see Sex cells, in female
Ejaculation, 123, 124, 152
Embryo, 134
Emergency action, 40, 41
Emotional disorders, 42
Emulsification, 104
Enamel, of tooth, 97
Endocardium, 75
Endocrine glands, 140–147, 152
    in male, 122
Energy, conservation of, 40
Enterokinase, 102
Enzymes, 95, 152
Epidermis, 59
    See also Skin
Epididymis, 123
Epinephrine, 146
    effects of, 146
Epiphyseal plate, 7
Equilibrium, 53, 54
Erection, 124
Erepsin, 102
Erythrocytes, 80, 81
Esophagus, 100
Estrogen, 132–133, 142
    in female, 126
Eunuch, 125
Eustachian tube, 54
Excreta, 113

160 INDEX

Excretion, 113, 152
   sweating, 62
Excretory system, 112–118
Exercise, 18
Exocrine glands, 140, 153

Expiration, see Breathing
Exposure, to sun, 69
Extensor muscles, of forearm, 24
External anal sphincter, 25
External ear, see Ear, parts of
External respiration, 87
Extrinsic ocular muscles, 19
Eye, care of, 51–53
   color of, see Iris
   cross section of, 46
   disorders of, 48–51
   examination of, see Optometrist
   exercises for, 52
   injuries, 52, 53
   irritations, 52, 53
   pupil of, 39
   structure of, 45–48
Eye doctor, see Ophthalmologist
Eye surgery, see Ophthalmologist
Eyeglasses, 50
Eyestrain, 48, 52, 53

Face, 3, 4
Fallopian tubes, 126–127
Farsightedness, see Hyperopia
Fat, metabolism of, 104
   storage of, 14
Fatigue, 15, 18
   and proper lighting, 53
Fear, 40
Feces, color of, 104
Femur (thigh bone), 6, 25, 26
Fertilization, 124
   exact day of, 135
   period of, 127
Fetal circulation, 80
Fetus, 153
Fibrinogen, 82
Fibrocartilaginous callus, 10
Fibula, 6
Filtration, 117
      See also Urine
Fingers, 6
Flat bones, 3
Flexor carpi radialis, 24
Flexor carpi ulnaris, 24
Floating kidney, 114
Focal point, 48
Focus, 45, 46
Follicle-stimulating hormone (FSH), 125, 132–133, 141
Fontanels, 4
Food, absorption of, 95
Food tube, see Esophagus
Foot, bones of, 6
   muscles of, 27
Foramen magnum, 3
Forearm, muscles of, 24

Foreskin, of penis, 124
      See also Circumcision
Fovea, 48
Fracture, of bone, 7–11
Fractures, complete, 9
   compound, 9–10
   healing of, 9–11
   incomplete, 9
   simple, 10
Frontal bone, 3
Frontal lobe, 36
FSH, see Follicle-stimulating hormone
Functional disorders, see Emotional disorders
Fundus, 100
Fungi, 66

Gallbladder, 104
Gastric digestion, 95, 100
Gastric juice, 100, 101
Gastric lipase, 101
Gastric protease, see Pepsin
Gastrocnemius, 26
Genetic, 153
Genetic material, of cell, 33
Genitalia, external, 128
GH, see Growth hormone
Giantism, 141
Gland, definition of, 139
      See also Cowper's glands, Ovaries, Pituitary gland, Prostate gland, Seminal vesicles, Testes
Glandular secretion, 40
Glans penis, 124
Glaucoma, 51, 153
Gliding joints, 9
Glomerulus, 116
Glucagon, 147
Glucose, 147
   absorption of, 104
Gluteus maximus, 25
Gluteus medius, 25
Glycogen, 18, 147
Glycosuria, 118, 147, 153
Goggles, 53
Goiter, 144
Gonadotropins, 125, 131, 141–142
      See also FSH, ICSH, LH, LTH
Gonads, in female, see Ovaries
   in male, see Testes
   secretions from, 147
Gonatropic, 153
Goose flesh, 63
Graafian follicle, 126
Gracilis muscle, 26
Gray matter, in brain, 36, 37
Gristle, 153
Growth hormone (GH), 141

Hair, 62–63
   care of, 70–71
   cycles, 63

INDEX

follicle, 60, 63
   root, 63
Hairsprays, and the eye, 52
Hammer (malleus), 54
Hamstring muscles, *see* Biceps femoris
Hand, bones of, 6
   muscles of, 24
Hangnails, 63
Headaches, and proper lighting, 53
Hearing, 53–54
Hearings aids, 55
Heart, 74–78
   chambers of, *see* Atrium and Ventricle
   physiology of, 77
Heart murmur, 78
Heartbeat, 37, 40
   per minute, 77
Heat, as waste, 113
Hematuria, 118, 153
Hemocytometer, *see* Blood count
Hemoglobin, 80
   and skin color, 61
Henle's loop, 115–117
Heredity, and baldness, 71
   and color blindness, 50
   and eye disorders, 48
   and skin color, 60–61
   and tooth development, 97
Hilum, 114
Hinge joints, 9
Hip and buttock, muscles of, 25
Histamine, 70
Hives, *see* Allergies
Hormones, 140, 153
   reproductive, 131
Hot flashes, and menopause, 131
Humerus (upper arm), 6, 23
Hydrochloric acid, 100
Hymen, 128
Hyperglycemia, 147
Hyperinsulinism, 147
Hyperopia (farsightedness), 48–49, 153
Hyperparathyroidism, 145
Hyperthyroidism, 144
Hypoglycemia, symptoms of, 147
Hypoinsulinism, 147
Hypoparathyroidism, 145
Hypophysis, *see* Pituitary gland
Hypothyroidism, 144
Hysterectomy, 131

ICSH, *see* Interstitial-cell-stimulating hormone
Ileocecal valve, 105
Ileum, 102
Ilium, 26
Immature follicle, 126
Impetigo, 65–66
Impotence, 124, 153
Incisor teeth, 97
Indigestion, 41
Infection, and baldness, 71
   of skin, 65
Infertility, 122
Insecticides, 35
Inspiration, *see* Breathing
Insulin, 104, 147
Intercostals, 23
Intermediate lobe, of pituitary gland, 143
Internal ear, *see* Ear, parts of
Internal respiration, 87
Interstitial-cell-stimulating hormone (ICSH), 125, 142
Intestinal absorption, 104
Intestinal digestion, 95, 102
Intestinal juice, 104
Intestinal lipase, 102
Intestine, and ulcers, 41
   See also Large intestine, Small intestine
Intramuscular, 153
Iodine, 144
Iris, 46
Irregular bones, 3
Islets of Langerhans, 104, 147
Isometric contractions, 18
Isotonic contractions, 18

Jaundice, 83
Jejunum, 102
Joints, 9

Keratin, 60, 153
Kidney stones, 118, 145
Kidneys, 113
   structure of, 114
Knee cap, *see* Patella

Labia, majora and minora, 128
Lactase, 102
Lactation, 143
Lacteal, 104
Large intestine, 105, 108
Larynx, 89
Lateral, 153
Lateral abdominal muscles, 25
Latissimus dorsi, 23
Learning, 41
Leg, muscles of, 26
   See also Thigh
Lens, *see* Crystalline lens
Lenses, prescription, 52
   See also Optician
Leukocytes, 80
   and infection, 81
Levator ani, 25
LH, *see* Luteinizing hormone
Lice, 67
Ligaments (connective tissues), 1, 9
Light rays, *see* Light waves
Light waves, 45–46
   See also Ultraviolet rays
Lighting, proper, 53
Little brain, *see* Cerebellum
Liver, function of, 104

Lockjaw, see Tetanus, disease of
Long bones, 3
LTH, see Luteotropin
Lumbar vertebrae, 23
Lumen, 102, 104
Lungs, 89
   capacity of, 90–91
   and the circulatory system, 76
Lunule, 63
Luteinizing hormone (LH), 132–133, 142
Luteotropin (LTH), 132–133, 142
Lymph, 78
Lymph glands, see Lymph nodes
Lymph nodes, 79
Lymphatic system, 78–79
Lymphocytes, 79, 81

Maltase, 102
Mammary glands, 142
Massage, 19
Masseter, 19
Mastication (chewing), 19, 97
Mastoid process, 20
Measles, and hearing loss, 55
Medulla, 153
   of adrenal gland, 145–146
   of kidney, 114
Melanin, 60, 62, 153
   and birthmarks, 69
Melanocyte-stimulating hormone (MSH), 143
Melanocytes, 60
Memory, 36, 41
Menarche, 129, 153
Meninges, 37
Menopause, 129, 131, 153
Menorrhagia, 129
Menstrual cycle, 128, 130, 134
   duration of, 129
Menstrual period, duration of, 129
Menstruation, 129, 153
   disturbances of, 129–131
Mental activity, and fatigue, 19
Metabolism, 144
Metacarpals, 6
Metatarsals, 6
Microorganisms, as waste, 113
Middle ear, see Ear, parts of
   bones of, see Hammer, Anvil, Stirrup
Milk production, see Lactation
Minimal air, 91
Miscarriage, 133, 135
Mites, see Scabies
Mitral valve, 75, 77
Molar teeth, 97
Molds, see Fungi
Moles, 69
Monocytes, 81
Monosaccharides, see Sugar
Mouth, see Oral cavity
Movement, and joints, 9
MSH, see Melanocyte-stimulating hormone
Mucin, 99

Mumps, and hearing loss, 55
Muscle, belly of, 14
   definition of, 13
   insertion of, 8, 14
   origin of, 8, 14
Muscle cells, size of, 15
Muscle contraction, and body temperature, 16
   chemistry of, 18
   latent period of, 16
   maximum, 16
   relaxation period of, 16
   summation of, 17
   types of, 18
Muscle fibers, 14
Muscle tissue, kinds of, 13
Muscles, of the head, 19
Muscular system, 13–27
Myocardium, see Cardiac muscle
Myofibrils, 15
Myopia (nearsightedness), 48, 49, 153
Myxedema, 144

Nails, structure of, 63
Nasal cavity, 48
Nasal chamber, 89
Nearsightedness, see Myopia
Neck, muscles of, 20
Neck, of tooth, 98
Nephron, 115
   function of, 116
Nerve, 33
Nerve fibers, 33
   tracts of, 36
Nerve gases, 35
Nerve impulses, transmission of, 34
Nerves, cranial, 38
   spinal, 38
   vagus, 38
Nervous system, autonomic, 39–41
   central, 35–38
   disorders of, 42
   peripheral, 38–41
   structure of, 32
Neurological disorders, 42
Neurologists, 42
Neuron, 32, 33
Neurons, 39, 41
   connector, 33
   motor, 33
   sensory, 33
Niacin (nicotinic acid),
   deficiency symptoms, 107
   function of, 107
   properties of, 107
   sources of, 107
Nits (louse eggs), see Lice
Noradrenalin, see Norepinephrine
Norepinephrine, 146
Nose, see Nasal chamber
Nucleus, of cell, 33
Nutrients, 18
   for joints, 9

Occipital bone, 3

Occipital lobe, 36
Oil glands, 60
Ophthalmologist, 50, 51, 52
Optic nerve, 47
Optician, 51
Optometrist, 51, 52
Oral cavity, 97
Oral temperature, 61
Orbicularis oculi, 19
Orbicularis oris, 19
Organic disorders, see Neurological disorders
Orgasm, in male, 125
Orthodontia, 98
Ossicles, 54
Ossification, 6, 9
Otologist, 54
Ova, see Sex cells, in female
Ovaries, 125
Oviducts, see Fallopian tubes
Ovulation, 127
Oxygen, 18, 86
  in blood, 91
  shortage of, 41
  transport of, 91–92
Oxygen debt, 18
Oxyhemoglobin, 92
Oxytocin, 143

Pacemaker, see Sinoatrial node
Palate, 97
Pancreas, 147
  function of, 102, 104
Pancreatic amylase, 101
Pancreatic lipase, 101
Pancreatic protease, 101
Parasympathetic division of nervous system, 40
Parathormone (PTH), 145
Parathyroids, 144–145
Parietal bones, 3
Parietal lobe, 36
Patella (knee cap), 6, 25
Pathogens, 81
PBI, see Protein-bound iodine
Pectoral girdle, 4
Pectoralis, major, 23
  minor, 23
Pelvic girdle, 6
Pelvic floor, muscles of, 25
Penis, 123, 124
Pepsin, 100, 101
Peptic ulcers, see Ulcers, digestive
Pericardial sac, 75
Period, see Menstrual period
Periodontal membrane, 98
Periosteum, 7
Peripheral nervous system, see Nervous system
Peristalsis, 100
Permanent teeth, 97
Peroneus longus and brevis, 26
Phalanges, 6
Pharynx, 89, 97, 99, 100
Pigments, 60

Pimples, see Acne
Pinna, 53
Pituitary gland, 125, 140–143
  location of, 142
Plasma, 80
Platelets, 80, 81
Platysma, 20
Posterior abdominal muscles, 25
Posterior lobe, of pituitary gland, 143
Posture, 15, 154
Pregnancy, 134–135
  length of time, 135
  and menstruation, 129, 134
  and Rh factor, 83
Premature birth, 135
Prepuce, see Foreskin
Presbyopia, 50, 154
Procallus, formation of, 10
Progesterone, 132–133
  in female, 126
Prolactin, 133
Pronator teres, 24
Prostate gland, 123
Protein-bound iodine, 144
Proteins, metabolism of, 104
Proximal convoluted tubule, 115
Psychiatrists, 42
Psychologists, 42
Psychosomatic disorders, 40, 41
Psychotherapy, 42
PTH, see Parathormone
Ptyalin, see Salivary amylase
Puberty, 154
  in female, 126
  in male, 123
Pulmonary artery, 74, 76
Pulmonary valve, 76
Pulp cavity, of tooth, 98
Pulse, 78
Pupil, of eye, 46
Purkinje fibers, 77
Pus, 81
Pylorus, 100

Quadriceps femoris, 25

Radius, 6
  See also Forearm
Rashes, and cosmetics, 64, 65
  See also Skin, special problems of
Reabsorption, 117
Rectal temperature, 61
Rectum, 105
Red blood cells, see Erythrocytes
Reflex action, 39, 154
Regeneration, 121
Renal artery, 114
Renal pelvis, 114
Renal vein, 114
Renal pyramid, 114, 115
Renal sinus, 114
  See also Kidneys
Rennin, 101
Reproductive hormones, in male, 125

164 INDEX

Reproductive system, of female, 125, 134
  of male, 122–125
Residual air, 91
Respiration, and blood stream, see Internal respiration
Respiratory system, 86–92
  parts of, 88–89
Retina, of eye, 47, 48, 50
Rhesus protein, see Rh factor
Rheumatoid arthritis, 146
Rh factor, 83
Rho GAM, 83
Rhomboids, 23
Riboflavin (vitamin B_2)
  deficiency symptoms, 107
  function of, 107
  properties of, 107
  sources of, 107
Ribonucleic acid (RNA), 41
Ringworm, 66–67
RNA, see Ribonucleic acid
Rods, 47, 48
Root, of tooth, 97
Roughage, 108

SA node, see Sinoatrial node
Saccule, 54
Sacrospinalis, 23
Sacrum, 6, 23
Saliva, 97
Salivary amylase, 99, 101
Salivary digestion, 95, 98–100
Salivary glands, 98–99
Salts, absorption of, 104, 105
Sarcolemma, 15
Sarcoplasm, 15
Sartorius, 26
Scabies, 67
Scapulae (shoulder blades), 6, 23
Scarlet fever, and hearing loss, 55
Sclera, of eye, 46
Scrotum, 122
Sebaceous glands (oil glands), 63
Secondary sex characteristics, in female, 126
  in male, 123
Semen, 124
Semicircular canals, 54
Seminal ducts, 123
Seminal fluid, see Semen
Seminal vesicles, 123
Seminiferous tubules, 123
Septum, 77, 154
Serum, 80
7-dehydrocholesterol, see Vitamin D, production of
Seven-year-itch, 67
Sex cells, in female (eggs), 125
Sex hormones, in male, 122
Sexual activity, and acne, 68
Sexual intercourse, 124, 135
  and lice, 67

SH (somatotropic hormone), see Growth hormone
Shampooing, 70
Shin bone, see Tibia
Short bones, 3
Shoulder, muscles of, 23
Shoulder blades, see Scapulae
Sinoatrial node (SA node), 77
Skeletal muscles, 13
  contraction of, 16
  kinds of, 19
  number of, 19
  stress on, 8
Skeletal system, 1, 2
Skin, care of, 63–65
  color of, 60
  functions of, 61
  pigmentation of, 143
  and sensory reception, 62
  special problems of, 65–70
  structure of, 59–61
Skin cancer, 69
Skin rashes, see Rashes
Skull, fetal development of, 3
  number of bones, 3
Small intestine, parts of, 102
  size of, 102
Smooth muscle, 13
Soleus, 26
Somatotropic hormone (SH), see Growth hormone
Sound waves, 54
Sperm, production of, 122
Sperm count, 125
Sperm cells, parts of, 123
  size of, 123
Sphincter muscles, 105
Sphincter urethrae, 25
Sphygmomanometer, 154
  See also Blood pressure, measurement of
Spinal canal, 37
Spinal column, 4
Spinal cord, 35, 37
Spinal nerves, see Nerves, spinal
Spontaneous fracture, 9
Spotting, in menstruation, 129
Staphylococcus aureus, 65
Starch, 99
  See also, Glycogen
Steapsin, see Pancreatic lipase
Sterility, 122, 124
Sternocleidomastoid, 20
Sternum (breastbone), 4, 20, 23
Stethoscope, 78
Stillbirth, 135
Stimulus, 39
  of muscle, 15
Stirrup (stapes), 54
Stomach, capacity of, 100
Stomach acid, 41
Strabismus, 50, 154
Stress, 40, 41, 42, 146

INDEX

Subconscious, 40
Sucrase, 102
Sugar, 99, 105
Summation of contractions, *see* Muscle contraction, summation of
Sunburns, 68, 69
Sunglasses, 52
Suntans, 68, 69
Supplemental air, 90, 91
Suspensory ligament, of eye, 46
Sutures, 3, 9
Sweat, 62
Sweat glands, 60
  and deodorants, 64
Sympathetic division of nervous system, 40
Symphysis pubis, 6
Synapse, 34, 41
Systole, 77, 154

Tailor's muscle, *see* Sartorius
Tampons, 128
Tanning, 60
Tarsels, 6
Tear glands, 48
Teeth, function of, 97
  number of, 97
Temperature, *see* Body temperature
Temporal bones, 3
Temporal lobe, 36
Temporal muscle, 19
Temporary teeth, 97
Tendons, 8
Testes, 122, 142
Testicles, *see* Testes
Testosterone, 125
Tetanus, disease of, 18
  state of, 17, 18
Tetany, 145
Theory, 154
Thiamine (vitamin B_1),
  deficiency symptoms, 107
  function of, 107
  properties of, 107
  sources of, 107
Thigh, muscles of, 25, 26
  See also Femur
Thoracic vertebrae, 23
Thorax, muscles of, 23
Throat, *see* Pharynx
Thrombocytes, *see* Platelets
Thyroid gland, 143–145
  location of, 143
  weight of, 143
Thyroid-stimulating hormone (TSH), 143
Thyroxine, functions of, 144
Tibia (shin bone), 6
Tibialis anterior, 26
Tibialis posterior, 26
Tidal air, 90

Toe dancer's muscle, *see* Gastrocnemius
Toes, muscles of, 26
Tone, 15
  in abdominal wall, 15
  in small arteries, 15
Tongue, extrinsic muscles of, 19
  intrinsic muscles of, 19
Tonometer, 51
Tonus, *see* Tone
Tooth, structure of 97–98
Trachea, 89
Trapezius, 20, 23
Traumatic fractures, 9
Triceps brachii, 24
Tricuspid valve, 75, 77
Triiodothyronine, 144
Tropins, 141
Trypsin, *see* Pancreatic protease
TSH, *see* Thyroid-stimulating hormone
Tumors, and menstruation, 129
Twitch, 17
Tympanic membrane (eardrum), 53–54

Ulcers, digestive, 41
Ulna (forearm), 6
Ultraviolet rays, 52
  and skin color, 61
Umbilical cord, 80
Unconscious, 154
Universal acceptors, *see* Blood, transfusions
Universal donors, *see* Blood, transfusions
Upper arm, *see* Humerus
Upper leg, *see* Femur
Ureter, 113, 114, 116
Urethra, 113
  in female, 116
  function of, 116
  in male, 116, 123
Urinary bladder, 113, 116
  capacity of, 116
  location of, 116
Urinary system, parts of, 113
Urination, 118
  abnormal conditions of, 118
Urine, 113
  color of, 118
  composition of, 117–118
  formation of, 117
  volume of, 117–118
Uterus, 126–127
  location of, 127
  size of, 127
Utricle, 54

Vagina, 128
Vagus nerves, *see* Nerves, vagus
Vas deferens, *see* Seminal ducts
Vasopressin, *see* ADH
Vein, 76, 154
Ventral root, 38

Ventricle, of heart, 75
Venules, 76
Vermiform appendix, see Appendix
Vertebrae, 4, 37
Vertebral column, 4, 5
  number of bones, 4
  parts of, 4
Villi, 102, 103
Virginity, 128
Vision, 48
  black and white, see Rods
  blurry, 45, 48
  color, see Cones
  peripheral, 51
Vitamin A_1,
  absorption of, 105
  deficiency symptoms, 106
  function of, 106
  properties of, 106
  sources of, 106
Vitamin B_1, see Thiamine
Vitamin B_2, see Riboflavin
Vitamin B_6,
  deficiency symptoms, 107
  function of, 107
  properties of, 107
  sources of, 107
Vitamin B_{12}, see Cyanocobalamin
Vitamin C (ascorbic acid),
  absorption of, 105
  deficiency symptoms, 107
  function of, 107
  properties of, 107
  sources of, 107
Vitamin D,
  absorption of, 105
  deficiency symptoms, 106
  function of, 106
  production of, 62
  properties of, 106
  sources of, 106
Vitamin E,
  deficiency symptoms, 106
  properties of, 106
  sources of, 106
Vitamin K,
  absorption of, 108
  deficiency symptoms, 106
  function of, 106
  properties of, 106
  sources of, 106
Vitamins, absorption of, 105
Vital capacity of lungs, 91
Vital organ, 154
Vocal cords, see Larynx
Voice box, see Larynx
Vulva, 128

Warts, 69
Waste removal, 105
Wastes, 18
  from joints, 9
Water, absorption of, 104–105, 108
  as waste, 113
Waves, light, see Light waves
Waves, sound, see Sound waves
White blood cells, 78–79
      See also Leukocytes
White matter, of brain, 37
Windpipe, See Trachea
Wisdom teeth, see Molar teeth
Womb, see Uterus
Wrist, 6